A NATURALIST'S GUIDE TO THE

LIZARDS
OF
AUSTRALIA

Scott Eipper & Tyese Eipper

JOHN BEAUFOY PUBLISHING

To Hal and Heather, your work has been an inspiration to herpetologists across the globe. This book is a small gesture of our appreciation for your efforts. Thank you.

First published in the United Kingdom and Australia in 2021 by John Beaufoy Publishing Ltd
11 Blenheim Court, 316 Woodstock Road, Oxford OX2 7NS, England www.johnbeaufoy.com
Copyright © 2021 John Beaufoy Publishing Limited
Copyright in text © 2021 Scott Christopher Eipper and Tyese Eipper
Copyright in photographs © see below and as credited throughout
Copyright in maps © 2021 John Beaufoy Publishing Limited

Photo Credits
Front cover: Central Bearded Dragon, S. Eipper. **Bottom row, left to right** Emerald Monitor, T. Eipper; Leopard Ctenotus, S. Eipper; Angle-headed Dragon, S. Eipper.
Back cover: Common Scalyfoot, T. Eipper.
Title page: Perentie, S. Eipper.
Contents page: Beautiful Velvet Gecko, S. Eipper

ISBN 978-1-913679-06-4

Edited by Krystyna Mayer

Designed by Gulmohur Press, New Delhi

Printed and bound in Malaysia by Times Offset (M) Sdn. Bhd.

·CONTENTS·

INTRODUCTION

Australia is currently home to 856 species and 59 subspecies of lizard, 830 of which are endemic. Many are being investigated and inevitably more will be described. Since 2018, over 50 Australian lizard species have been described or resurrected from synonymy. There are six new invaders – four house gecko species *Hemidactylus* spp., the Mourning Gecko *(Lepidodactylus lugubris)* and the skink *Lygosoma bowringii* are now established. Conversely, one species – the Garden Skink – is established overseas in New Zealand and Hawaii.

LIZARD FAMILIES

This book focuses on the classification of reptiles below the taxonomic rank of order Squamata (lizards and snakes), suborder Lacertilia (lizards). Lizards are further divided worldwide into 27 families. Australia is home to seven of them: the Gekkonidae, Carphodactylidae and Diplodactylidae (geckos), Pygopodidae (flap-footed lizards), Scincidae (skinks), Agamidae (dragons), and Varanidae (monitor lizards). A brief overview of each Australian family is provided below.

Australian Geckos

These belong to three families: cosmopolitan geckos (family Gekkonidae), carphodactylid geckos (family Carphodactylidae) and diplodactylid geckos (family Diplodactylidae). All three families are found throughout the mainland apart from the far south-east and Tasmania, and 208 species are currently described and accepted, in 24 genera. Geckos are

T Eipper

Rough-throated Leaf-tailed Gecko

4

nocturnal and eat mainly invertebrates such as spiders and insects, though some species also eat small lizards. Most geckos, like flap-footed lizards and skinks, can drop their tails. This strategy is used when a lizard is threatened, in the hope that a predator will pay more attention to the tail writhing in front of it, while the lizard makes a hasty retreat. As lizards use their tails for fat storage the strategy is thought to be used as a last resort. The tail does regrow, but the new tail is different in colour, pattern and shape, and often without tubercles. If startled, fighting or displaying courtship behaviour, almost all geckos can emit noise, either a barking or a squeaking sound. Australian geckos do not have moveable eyelids, and their eyes are covered by a transparent spectacle that comes off when they slough.

All **carphodactylid geckos** are endemic to Australia. They have notable tails – some are broad and flat, some have terminal knobs and others contain spikes. They lay two soft-shelled eggs per clutch. **Diplodactylid geckos** occur in Australia, New Zealand and New Caledonia, and most have padded digits. They account for the majority of Australian geckos. All Australian diplodactylid geckos are oviparous, usually laying two soft-shelled eggs per clutch, although those found in New Zealand and New Caledonia are viviparous, usually producing twins. **Gekkonid geckos** are found worldwide. They differ from other Australian geckos by producing hard-shelled eggs.

Flap-footed Lizards

These lizards in the family Pygopodidae are endemic to Australasia and are closely related to diplodactyline geckos; 46 species are currently described and accepted, in seven genera. Pygopods have no forelimb and the hindlimb is a flap slightly above the vent. Their eyes are like those of geckos, without a lid and covered by a transparent spectacle. All are oviparous and breed in spring, laying 1–2 long, soft-shelled eggs. They are cathemeral, feeding mainly on invertebrates, although some species also eat small skinks.

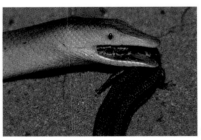

Above and below, left and right: *Burton's Legless Lizard capturing and swallowing a Boulenger's Skink*

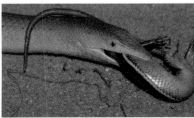

Australian Skinks

These lizards are the epitome of diversity. There are four subfamilies in Australia and its territories: the Egerniinae, Eugongylinae, Lygosominae and Sphenomorphinae. It is possible that further work may elevate these groups to family level, and 460 species are currently described and accepted, in 46 genera. Their diversity has allowed skinks to occupy every ecological niche of Australia, excluding marine and Subantarctic conditions. From the giant smooth-scaled skinks in the genus *Bellatorias* to the diminutive *Pygmaeascincus*, there are skinks to suit all environments. Some species have exploited

S Eipper

Shark Bay Shingleback

urban environments, like the fence lizards *Cryptoblepharus* running over houses, while blue-tongues *Tiliqua* steal bites from fruits and vegetables and make excellent pets. The legless varieties are often mistaken for small snakes and sadly many are killed due to this similarity. Unfortunately, one species of Australian skink, the Christmas Island Forest Skink *Emoia nativitatis*, became extinct in 2014 due primarily to invasive exotic species.

Dragons

The dragons belong in the Agamidae family, with 107 species currently described and accepted, in 16 genera. For the most part they live in open, dry habitats, although some species have adapted to live beside waterways and in forests. Their incredible scalation is the most diverse of all Australian lizards. From the vertebral crests and spine-like projections on the forest dragons, to the rock-mimicking earless dragons, their ecology is as varied as their habitat choice. Australian dragons utilize some of the most amazing anti-predator strategies in all reptiles, such as the extended spikes of the Thorny Devil and the erectable frill of the Frilled Dragon, and these iconic species are known by reptile enthusiasts across the globe.

A Elliott

A pair of Peninsula Dragons, female on left, male on right

One feature that differentiates dragons from all other Australian lizards is the tongue – it is small and thick, used only for picking up food and pheromone transport into the Jacobson's organ. All Australian dragons are oviparous, laying soft-shelled eggs. They are predominantly diurnal, feeding mainly on invertebrates, although some eat vegetation and small vertebrates, and others are more specialized, eating a specific genus of ant. Unfortunately, one species of Australian dragon, the Victorian Earless Dragon *Tympanocryptis pinguicolla*, is thought to have become extinct in 1969 due to invasive species and habitat destruction.

Monitors or Goannas

These lizards in the Varanidae family can be found throughout mainland Australia, and 30 Australian species are currently recognized and accepted, in a single genus. Monitors are carnivorous, diurnal predators. Some large species are scavengers, relying on their chemoreceptors to find their next meal. All monitors are egg layers, laying their soft-shelled eggs in nests, burrows dug into the ground or in termite mounds. They are unable to regrow their tails. However, their tails have been adapted for their lifestyle – a laterally compressed tail helps an aquatic monitor to swim, a prehensile tail is perfect for climbing rainforest trees, and a spiny tail can block entrances to burrows or wedge itself between rocks. The muscular tail can also be used like a whip, along with loud hissing and gaping if threatened. As a last line of defence, some species lunge and bite an aggressor.

Lace Monitor hatching from its egg

S Eipper

LIZARD FACTS

Lizard Myths

- Blue tongues keep snakes away from lizards.
- Monitor bites reopen every year. Monitors, like most animals, can bite humans if provoked, and their bites can become infected but do not reopen.
- The forked tongue is a sting.
- The frill of a Frilled Lizard is used as a parachute/flying aid to help it descend from trees. This was thought to be the case many years ago but was proven to be incorrect.
- Lizards are attracted to saucers of milk. Many animals are lactose intolerant and milk is not a part of lizards' normal diet.

Lizard Bites & First Aid

Almost all lizards are essentially harmless, with the exception of large monitors, which can have very sharp teeth that are recurved, with a slicing edge designed to cut through the flesh of prey. The bite may cause significant injury to a human, requiring surgery if severe. Large dragons can have long, canine teeth and strong jaw musculature that can result in an injury. Large skinks can have strong jaw musculature and can crush digits, particularly in small children. Bites should be treated via typical first-aid procedures for an open wound, including cleaning of the bite site and potentially seeking professional medical care.

While some Australian lizards are venomous, the venom has not been demonstrated to cause any levels of significant harm. Typically, the venom results in increasing stinging around the bite site, itchiness and increased bleeding. These effects usually rapidly subside.

Lizard or Snake

• Lizards usually have eyelids, while snakes have a fixed scale called a spectacle or brille covering the eye. The exception to this are geckos, legless lizards and some skinks.

S Eipper

Striped Worm Lizard

• Lizards usually have ear openings, while snakes do not have ears and are fundamentally deaf to airborne sound.
• Most lizards have a broad, fleshy tongue. The exception to this are the varanoid lizards, such as monitors, which have long, snake-like tongues.
• Snakes do not have legs. Pythons and blind snakes have a vestigial pelvic girdle, represented externally by the cloacal spurs. Most lizards have limbs, but in some fossorial species and legless lizards these have become vestigial or non-existent.

Lizards Around the Home

Gardens and urban environments can provide suitable homes for many lizard species.

T Eipper

Eastern Water Dragons are commonly seen across eastern Australian cities

They are not pests and have excellent benefits for humans, from consuming insects and spiders, to retaining biodiversity in an urban environment. Lizards can be encouraged around the home by native gardens that have suitable refuges such as rocks and logs among plants.

Lizards are becoming increasingly popular as pets. While being quieter than *traditional* pets, they can have an expensive price tag to set up correctly and are a long-term commitment, with many having a lifespan of more than 20 years if cared for properly. In Australia there are laws pertaining to keeping native wildlife as pets, each state being different from the others. Once you have chosen the species you would like to keep, you must determine if you can provide its needs. Check with the appropriate government department before acquiring a lizard as a pet to make sure you have the correct licence. There are many good books covering lizard keeping, some of which are listed in the further reading section (p. 171).

Habitats

Australian lizards occur in a wide variety of habitats. Some species are generalists, while others are restricted to small microhabitats within larger ecological communities. A sample of the variety of important habitats to Australian herpetofauna is shown below.

Alpine herb field   *Alpine grassland*   *Black-soil plains*

Brigalow open woodland   *Sand ridges with spinifex*   *Mallee woodland*

Chenopod scrubland

Open woodland

Rainforest

Rainforest stream

Tropical woodland

Urban modified woodland

Rocky escarpments

Gibber desert

Mangrove

LIZARD IDENTIFICATION

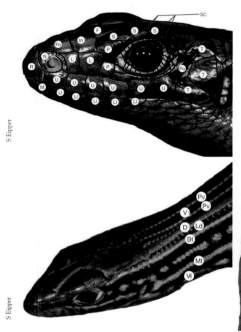

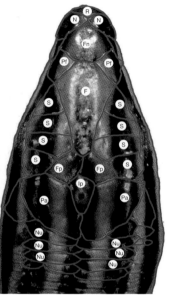

S Eipper

Scalation Key

R	Rostral	**S**	Supraocular	**Pf**	Prefrontal
N	Nasal	**Sc**	Supracillaries	**Dl**	Dorsolateral Stripe
F	Frontal	**T**	Temporal	**Ld**	Laterodorsal Stripe
Fn	Frontonasal	**M**	Mental	**Ml**	Midlateral Stripe
Fp	Frontoparietal	**U**	Upper labial	**Pv**	Paravertebral Stripe
L	Loreal		(Supralabial)	**V**	Vertebral Stripe
P	Preocular	**Ll**	Lower labial	**Vl**	Ventrolateral Stripe
Po	Postocular		(Infralabial)		
Pa	Parietal	**In**	Internasal		
Nu	Nuchal	**Ip**	Interparietal		

Using This Book

This book provides up-to-date information on about a third of the described Australian species, and acts as an introductory guide to help in their identification. The species have been selected based on how commonly encountered they are, their uniqueness and to provide a wide cross-section of Australian lizard fauna. Taxonomy follows Cogger 2018, apart from the addition of newly described taxa.

DISTRIBUTION KEY
GDR Great Dividing Range; **NSW** New South Wales; **NT** Northern Territory; **PNG** Papua New Guinea; **QLD** Queensland; **SA** South Australia; **TAS** Tasmania; **VIC** Victoria; **WA** Western Australia; **NP** National Park; **NR** Nature Reserve.

KEY FEATURES & MEASUREMENTS
SVL Snout to vent length
Sizes quoted in species descriptions for body measurements (where available) are average maximum sizes, but exceptions can occur. Breeding information (such as clutch/litter sizes), is taken from current literature and should be treated as an indicative value, as ongoing research can change the values provided.

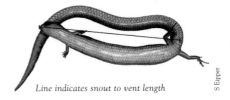

Line indicates snout to vent length

S.Eipper

Glossary

anterior Front of body; towards front.
aquatic Living in water.
arboreal Living predominantly above the ground, in vegetation or trees.
arid Refers to habitat subject to little or no rain.
autotomy Voluntary casting off of a lizard's tail as an anti-predator strategy. Tail breaks along specific fracture points and often moves to draw the attention of a predator.
basking Act of a lizard exposing itself to increased temperature in order to raise its core body temperature.
bimodal reproduction Very rare ability to be able to give birth to live young and lay eggs.
carrion Decaying flesh of dead animal.
cathemeral Active at any time of day or night.
caudal Of tail; at or near rear half of body.
chemoreceptor Organ that responds to chemical stimuli in the environment. Usually associated with senses of taste and smell.
chenopod shrubland Semi-arid plains vegetated with saltbush, samphire and similar.
clutch Number of eggs laid by female lizard in single reproductive event.
crepuscular Active at dawn and dusk.
cryptic Disguised appearance, through colour, body structure, pattern or habits.

diurnal Active during the daytime.

dorsal Of, on or relating to top or upper side of structure or body.

dorsolateral Region between top (dorsum) and sides (flanks) of a lizard.

dorsum Back of body.

family Taxonomic rank above genus and below order.

feral Refers to introduced animal that has become established in the wild.

flank Side of body.

fossorial Living on or active beneath the soil surface.

gene Basic unit of genetic control. Each gene has a specific function and is found on a specific section of a specific chromosome.

genus (plural **genera**) Taxonomic group above species and below family.

gravid Pregnant; abdominal cavity contains formed eggs or young.

heath Open uncultivated land, typically on acid sandy soil, with characteristic vegetation of low shrubs and coarse grasses.

herpetofauna Collective term referring to group of amphibians and reptiles.

hummock grassland Grassland occurring on infertile sand plains and dune fields, rocky hills and mountain range slopes, as well as normally dry watercourses and salt-lake systems.

hybrid Genetic combination as a result of mating of two different species or subspecies.

invertebrate Animal lacking backbone.

juvenile Young individual.

lateral Refers to areas of a lizard's body and tail between upperside (dorsal) and underside (ventral) parts of body (that is, its sides).

litter Number of young born by female lizard in single reproductive event.

longitudinal Extending along axis of body or direction from front to back or head to tail.

mallee Various shrubby, multiple-stemmed eucalyptuses found in semi-arid areas of southern Australia.

morphological Pertaining to form or structure of an animal, especially its external appearance.

mulga Wide range of arid and semi-arid vegetation communities, from low and open shrubland on stony soils, to open-forests 10 or more metres tall, which occupy deep soils on extensive red-earth plains. Generally dominated by Mulga *Acacia aneura*, but often co-occurs with eucalypts, other acacias, and various other arid shrubs and trees.

nape Back of neck.

nasal scale Scale that contains nostril.

neonate Newborn animal up to six weeks old.

nocturnal Active during the night.

nuchal Neck area following where head and neck join.

ocelli Dark edged, ring-like spot or blotch.

order Taxonomic rank above family and below class.

original tail Tail a lizard was born/hatched with.

oviparous Reproducing by laying eggs.

ovoviviparous Reproducing by forming unshelled egg sacs to house developing young, which are held inside female until ready to hatch, then expelled either still within egg sac or after they leave it.

paravertebral Region on either edge of spine.

posterior Rear of body; towards rear.

regenerated tail Tail that has been regrown following full or partial tail loss – often not as intricate as original tail.

rostral Scale on end of snout.

reticulation Body pattern created by pigment forming grid-like pattern

saxicoline Living on or around rocks.

species Basic unit of taxonomic classification.

species complex Group of animals composed of both described and undescribed taxa that are currently lumped under a single species name.

stippling Patterns or markings created by grouping together of numerous small dots.

subspecies Taxonomic category that is a variation in a primary (nominate) species brought about by geographical or genetic isolation, usually characterized by variation in morphological or genetic features.

supralabial scales Scales bordering upper lip from rostral to angle of upper and lower jaws.

SVL (snout–vent length) Refers to distance from tip of snout to transverse cloacal opening (vent) – used to determine size in reptiles, as many species can experience tail autonomy.

sympatry Refers to individuals that share same geographical area.

tail autonomy, see **autotomy.**

taxonomy Study of plants and animals leading to their description, classification and naming.

temporal Situated around temples.

terrestrial Living on or near the ground surface.

TL (total length) Total length of animal with original tail.

torpor Dormant period of physical inactivity, usually characterized by reduced body temperature and slowed metabolic rate.

tubercle Small, rounded, projecting part or outgrowth; knob-like process in skin or on a bone.

ventral Undersurface or belly of an animal.

vertebral Along line of spine.

vertebrates Animals with a backbone.

vestigial Refers to remnant of appendage or other structure that has lost its original purpose through evolution.

vine forest Closed forest with understorey of thick vines forming dense thickets.

viviparous Reproducing by giving birth to live young.

CARPHODACTYLID GECKOS

Chameleon Gecko ■ *Carphodactylus laevis* SVL 13cm

DESCRIPTION Body robust. Pale pinkish-brown to dark brown, with lighter underside. Original tail black with crisp white rings. Regenerated tail mottled brown. Toes have claws rather than pads. Strong vertebral ridge. **DISTRIBUTION** Wet Tropics region of north QLD, between Rossville and Cardwell.
HABITAT AND HABITS Restricted to rainforest. Shelters by day in leaf litter and beneath fallen epiphytic ferns. Nocturnal. Terrestrial. Sits upside down on small saplings about 30cm above the ground, where it ambushes passing invertebrates. Discarded regenerated tail makes loud squeaks while original tail does not.

S Eipper

Centralian Knob-tailed Gecko ■ *Nephrurus amyae* SVL 13cm

DESCRIPTION Australia's heaviest gecko. Body robust. Reddish-orange covered in small, creamy yellow, prominently raised rosettes. Head covered with fine black or brown reticulum. Underside white. Unable to discard tail. Toes have claws rather than pads.
DISTRIBUTION Southern NT and eastern WA, from the Macdonnell Ranges to Kutjuntari Rockhole, WA. **HABITAT AND HABITS** Restricted to rocky gorges and ranges on stony soils. Nocturnal. Terrestrial. Emerges from beneath rocks and timber to hunt for invertebrates. Tail taps on the ground when agitated.

S Eipper

Prickly Knob-tailed Gecko ■ *Nephrurus asper* SVL 12cm

DESCRIPTION Body robust. Pale brown to charcoal with or without pale bands; some individuals are reddish-orange. Head covered with fine black or brown reticulum. Underside white. Dorsum covered in lower tubercles. Unable to discard tail. Toes have

claws rather than pads. Probably a species complex. **DISTRIBUTION** QLD, from tip of Cape York south to Rockhampton, and west through interior to Windorah. **HABITAT AND HABITS** Found in drier habitats, from heaths, open woodland and brigalow, to rocky deserts. Nocturnal. Terrestrial. Emerges from beneath rocks and timber to hunt invertebrates. Tail taps on the ground when agitated.

Pale Knob-tailed Gecko ■ *Nephrurus laevissimus* SVL 10cm

DESCRIPTION Body robust. Pale pinkish-orange on body with pale grey head. Large, irregular dark grey to black blotches on head, nape and shoulders, as well as on hips and on to tail. Tail short, but can be discarded. Regenerated tail similar in colour to body but lacks tubercles. Body smooth. Toes have claws rather than pads. **DISTRIBUTION** Arid central Australia, from Tarcoola, SA, west to Wiluna, WA, north to southern Kimberleys, east in the NT, into the Tanami Desert to Errinundra. **HABITAT AND HABITS** Found in drier habitats in hummock grassland and sand-ridge deserts. Nocturnal. Terrestrial. Shelters beneath rocks and timber, and in burrows.

Smooth Knob-tailed Gecko ■ *Nephrurus levis* SVL 10cm

DESCRIPTION Body robust. Pale pinkish-brown to grey with three dark bands from nape to shoulder, covered in yellow to cream spots. Three pale bands – one on base of head, one on neck and one on shoulder region. Underside white. Dorsum covered in low, blunt tubercles. Original tail broad and can be discarded. Regenerated tail similar in colour to body but lacks tubercles. Toes have claws rather than pads. Two subspecies (*N. l.*

occidentalis and *N.l. pilbarensis*) currently recognized but may be elevated to full species due to their genetic divergence.
DISTRIBUTION Arid central and west Australia, from Port Hedland to Cunnamulla, QLD, south to Wentworth, NSW.
HABITAT AND HABITS Found in drier habitats, including mallee, heath, mulga, open woodland and deserts. Nocturnal. Terrestrial. Shelters beneath rocks and timber, and in burrows.

S Eipper

Starred Knob-tailed Gecko ■ *Nephrurus stellatus* SVL 9cm

DESCRIPTION Body robust. Pale pinkish-brown to grey with dark bands from nape to shoulder, covered in yellow to cream spots. Three pale bands just behind head. Underside white. Dorsum covered in low, blunt tubercles. Tail short and slender. Toes

have claws rather than pads. Probably a species complex. **DISTRIBUTION** Two separate populations, in southern WA around Southern Cross, and SA near Yalata to Whyalla.
HABITAT AND HABITS Found in drier habitats, from mallee, heath, mulga and open woodland, to deserts. Nocturnal. Terrestrial. Shelters beneath rocks and timber, and in burrows.

S Eipper

Southern Banded Knob-tailed Gecko
■ *Nephrurus wheeleri* SVL 10cm

DESCRIPTION Body robust. Pale pinkish-brown to reddish-brown with 4–5 dark brown or grey broad bands. Underside white. Dorsum covered in low, blunt tubercles. Original tail broad with low tubercles. Toes have claws rather than pads. The **Northern Banded**

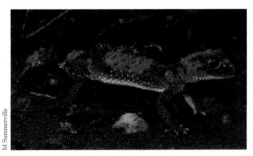

Knob-tailed Gecko *N. cinctus* was recently split off from this species. **DISTRIBUTION** Northern Goldfields and Murchison areas, WA. **HABITAT AND HABITS** Found in drier habitats, from mallee, mulga, open woodland and gorges, to rocky deserts. Nocturnal. Terrestrial. Emerges from beneath rocks and timber and from burrows to hunt invertebrates.

Mt Blackwood Broad-tailed Gecko ■ *Phyllurus isis* SVL 7.6cm

DESCRIPTION Body dorsoventrally compressed. Mottled greyish-brown, with lighter underside. Original tail has five white bands. Regenerated tail mottled brown. Body covered in lower tubercles and spines. Toes have claws rather than pads. **DISTRIBUTION** North QLD, on Mt Jukes and Mt Blackwood. **HABITAT AND HABITS** Restricted to rainforest and vine forest. Nocturnal. Emerges to sit upside down on rock faces, branches and vines near the ground, where it ambushes passing invertebrates.

Oakview Leaf-tailed Gecko ■ *Phyllurus kabikabi* SVL 8cm

DESCRIPTION Body dorsoventrally compressed. Mottled greyish-brown, with lighter underside. Original tail round and has 5–6 white bands. Regenerated tail mottled brown. Body covered in lower tubercles and spines. Toes have claws rather than pads. **DISTRIBUTION** South-east QLD, in small part of Oakview State Forest. **HABITAT AND HABITS** Restricted to rainforest and vine forest. Nocturnal. Emerges to sit upside down on rock faces and vines near the ground, where it ambushes passing invertebrates. One of Australia's most restricted reptiles.

S Eipper

Eungella Broad-tailed Gecko ■ *Phyllurus nepthys* SVL 10cm

DESCRIPTION Body dorsoventrally compressed. Mottled greyish-brown, with lighter underside with distinct dark flecks. Original tail flared, with four broken white bands. Regenerated tail mottled brown. Upper body covered in tubercles and long spines. Toes have claws rather than pads. **DISTRIBUTION** North QLD, near Eungella. **HABITAT AND HABITS** Restricted to rainforest and vine forest. Nocturnal. Emerges to sit upside down on rock faces, branches and vines near the ground, where it ambushes passing invertebrates. Also uses man-made structures such as buildings and bridges.

S Eipper

Broad-tailed Gecko ■ *Phyllurus platurus* SVL 10cm

DESCRIPTION Body dorsoventrally compressed. Mottled greyish-brown to yellowish with fine reddish markings; underside lighter. Original tail flared with tubercles. Regenerated tail mottled brown and smooth. Dorsum covered in tubercles. Toes have claws rather than pads.

DISTRIBUTION Eastern NSW, between Jamberoo and Denham. **HABITAT AND HABITS** Restricted to eucalypt forest and heath. Nocturnal. Emerges to sit upside down on rock faces, branches and vines near the ground, where it ambushes passing invertebrates. Also uses man-made structures such as buildings and bridges. Lays two eggs per clutch.

Northern Leaf-tailed Gecko ■ *Saltuarius cornutus* SVL 15cm

DESCRIPTION Body dorsoventrally compressed. Creamish-grey to brown, heavily mottled with brown and grey. Markings are like lichens on rainforest trees. Occasionally with a

greenish tint. Original tail has tubercules. Regenerated tail mottled brown and smooth. Upper body covered in tubercles. Toes have claws rather than pads. **DISTRIBUTION** North-east QLD, from Paluma to Rossville. **HABITAT AND HABITS** Restricted to rainforest and vine forest. Nocturnal. Emerges to sit upside down on trees, branches and vines near the ground, where it ambushes passing invertebrates. Also uses man-made structures such as buildings.

Rough-throated Leaf-tailed Gecko ■ *Saltuarius salebrosus* SVL 15cm

DESCRIPTION Body dorsoventrally compressed. Creamish-grey to brown, heavily mottled with brown and grey. Markings similar to lichens on rock faces. Occasionally with a greenish tint in rainforest-dwelling populations. Original tail has tubercules. Regenerated tail mottled brown and smooth. Upper body covered in tubercles. Toes have claws rather than pads. Probably a species complex. **DISTRIBUTION** QLD, from Blackdown NP to Bulburin NP, and inland to Carnarvon Gorge. **HABITAT AND HABITS** Restricted to rainforest, vine forest and eucalypt forest. Nocturnal. Emerges to sit upside down on trees and rock faces, where it ambushes passing invertebrates.

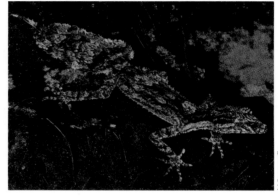

M Summerville

Southern Leaf-tailed Gecko ■ *Saltuarius swaini* SVL 14cm

DESCRIPTION Body dorsoventrally compressed. Creamish-grey to brown, heavily mottled with brown and grey. Markings similar to lichens on rainforest trees. Occasionally with a greenish tint. Original tail has tubercules. Regenerated tail mottled brown and smooth. Upper body covered in tubercles. Toes have claws rather than pads. **DISTRIBUTION** Eastern NSW and QLD, from Mt Glorious to Woodburn. **HABITAT AND HABITS** Restricted to rainforest and wet eucalypt forest. Nocturnal. Emerges to sit upside down on trees, branches and vines near the ground, where it ambushes passing invertebrates. Also uses man-made structures such as buildings.

S Eipper

Granite Belt Leaf-tailed Gecko ■ *Saltuarius wyberba* SVL 12cm

DESCRIPTION Body dorsoventrally compressed. Creamish-grey to brown, heavily mottled with brown and grey. Markings similar to lichens on trees. Original tail has tubercules.

Regenerated tail mottled brown and smooth. Upper body covered in tubercles. Toes have claws rather than pads. Probably a species complex. **DISTRIBUTION** QLD, from Queen Mary Falls to Tenterfield, NSW. **HABITAT AND HABITS** Restricted to heaths and eucalypt forest. Nocturnal. Emerges to sit upside down on rock faces and trees, where it ambushes passing invertebrates. Also uses man-made structures such as buildings.

Common Thick-tailed Gecko ■ *Underwoodisaurus milii* SVL 14cm

DESCRIPTION Body robust. Reddish-orange to purplish-brown with scattered white to yellow tubercles. In western animals tubercles align to form irregular bands. Original tail greyish-black with 4–5 white bands with low smooth tubercles. Regenerated tail mottled grey-brown and smooth. Toes have claws rather than pads. Probably a species complex. **DISTRIBUTION** Southern Australia, from Shark Bay, through WA, SA, VIC, ACT and NSW, to Rockhampton, QLD. **HABITAT AND HABITS** Widespread, using heaths, grassland, mallee, mulga and dry eucalypt forest. Nocturnal. Terrestrial. Emerges from beneath rocks and timber to hunt invertebrates.

Western population   *Eastern population*

Granite Belt Thick-tailed Gecko ■ *Uvidicolus sphyrurus* SVL 9cm

DESCRIPTION Body robust. Creamish-grey to pinkish-brown, flecked with black, brown and grey. Original tail greyish-black with 4–5 white bands with low, smooth tubercles. Regenerated tail mottled brown and smooth. Toes have claws rather than pads. **DISTRIBUTION** QLD, from Gatton, QLD, to Tamworth, NSW. **HABITAT AND HABITS** Restricted to heaths and open eucalypt forest with large rocks and boulders. Nocturnal. Emerges from beneath rocks to hunt invertebrates.

S Eipper

DIPLODACTYLID GECKOS

Clouded Gecko ■ *Amalosia jacovae* SVL 7cm

DESCRIPTION Body dorsally compressed. Light grey to black. Series of pale blotches forming broad, irregular vertebral stripe. Often light flecking over most of body. Pattern extends on to original tail. Regenerated tail same colour as body, with dark mottling. Underside white. Colour changes from day to night. Darker by day, with patterning less prominent. **DISTRIBUTION** South-east QLD, from Kroombit Tops to Warwick. **HABITAT AND HABITS** Restricted to heaths and eucalypt forest. Nocturnal. Often seen on trunks of smooth-barked trees and occasionally on houses.

S Eipper

Leseuer's Velvet Gecko ■ *Amalosia leseuerii* SVL 8cm

DESCRIPTION Body dorsally compressed. Light grey to black. Broad, dark-edged, irregular pale vertebral stripe. Light flecking often present over most of body. Pattern extends on to original tail. Regenerated tail same colour as body, with dark mottling. Underside white.

Colour changes from day to night. Darker by day, with patterning less prominent. **DISTRIBUTION** Southeast QLD, from Girraween, south into NSW to Nowra. **HABITAT AND HABITS** Restricted to heaths and eucalypt forest; usually found on rocky outcrops. Nocturnal. Often seen under rocks by day or hunting on rocks and nearby vegetation at night.

Zig Zag Velvet Gecko ■ *Amalosia rhombifer* SVL 7cm

DESCRIPTION Body dorsally compressed. Light grey to black. Broad, dark-edged, zigzag-shaped pale vertebral stripe. Pattern extends on to original tail. Regenerated tail same colour as body, with dark mottling. Underside white. Colour changes from day to

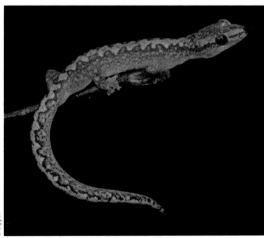

night. Darker by day, with patterning less prominent. Probably a species complex. **DISTRIBUTION** Northern and central Australia, from Eidsvold, QLD, to the Kimberley region of WA. Small population in NSW currently assigned to this species probably represents an undescribed taxon. **HABITAT AND HABITS** Restricted to dry open forest and creek lines, and usually found on rocky outcrops. Nocturnal. Often seen under rocks by day or hunting on rocks and trees at night.

Centralian Uplands Clawless Gecko
■ *Crenodactylus horni* SVL 3.5cm

DESCRIPTION Body lightly built. Light grey to brown. Pattern consists of broad stripes with fine flecking of lighter pigment. Pattern extends on to original tail. Regenerated tail same colour as body, with dark mottling. Underside white. Colour changes from day to night. Darker by day, with patterning less prominent. Probably a species complex. **DISTRIBUTION** Three populations, two near Alice Springs, NT. Third small population in south-west QLD currently assigned to this species probably represents an undescribed taxon. **HABITAT AND HABITS** Restricted to rocky outcrops with spinifex. Nocturnal. Usually seen on or near spinifex clumps hunting invertebrates. Lays 1–2 eggs per clutch.

H Cogger

Eastern Deserts Fat-tailed Gecko ■ *Diplodactylus ameyi* SVL 6cm

DESCRIPTION Body robust. Light grey to brown. Pattern of spots and fine flecks of light and dark pigment. Usually a pair of pale stripes from rostral along lateral sides of head. Pattern extends on to original granular tail. Regenerated tail same colour as body, with dark mottling, and smooth. Underside white. **DISTRIBUTION** Western QLD and NSW, from Winton to Peery Lake. **HABITAT AND HABITS** Found in stony soils, cracking clay grassland and open woodland. Nocturnal. Shelters in soil cracks, spider burrows and under rocks by day.

S Eipper

Mesa Gecko ■ *Diplodactylus galeatus* SVL 5.5cm
(Helmeted Gecko)

DESCRIPTION Body robust. Light reddish-brown. Pattern consists of 6–9 broad

blotches centred along spine. Head has pale blotch edged with black. Pattern extends on to original tail. Regenerated tail same colour as body, with dark mottling. Underside white. **DISTRIBUTION** Central Australia, from Alice Springs, NT, to Mt Eba, SA. **HABITAT AND HABITS** Restricted to rocky outcrops and gorges. Nocturnal. Shelters beneath stones and in spider burrows.

Kluge's Gecko ■ *Diplodactylus klugei* SVL 5.5cm

DESCRIPTION Body medium in build. Light reddish-brown. Pattern consists of broad, dark-edged blotches centred along spine. Pattern extends on to original tail. Regenerated tail same colour as body, with dark mottling. Underside white. **DISTRIBUTION** Exmouth to Murchison, WA. **HABITAT AND HABITS** Restricted to arid woodland and sand-ridge hummock grassland. Nocturnal. Shelters beneath rocks and fallen timber; also uses spider burrows.

Eastern Fat-tailed Gecko ■ *Diplodactylus platyurus* SVL 7cm

DESCRIPTION Body robust. Light grey to brown. Pattern of spots and fine flecks of light and dark pigment. Pattern extends on to original granular tail. Regenerated tail

same colour as body, with dark mottling, and smooth. Underside white. **DISTRIBUTION** Eastern QLD, from Blackdown Tableland to near Weipa on the Cape York Peninsula. **HABITAT AND HABITS** Found in stony soils, cracking clay shrubland and open woodland. Nocturnal. Shelters in soil cracks and spider burrows and under rocks by day.

Pretty Gecko ■ *Diplodactylus pulcher* SVL 6cm

DESCRIPTION Body medium in build. Light reddish-brown to brown. Pattern variable, consisting of broad, dark-edged blotches centred along spine to dark-edged stripe. Pattern

extends on to original tail. Regenerated tail same colour as body, with dark mottling. Underside white. **DISTRIBUTION** Perth to Onslow, WA, across to Yalata, SA. **HABITAT AND HABITS** Uses a wide variety of habitats, from dry woodland, to mulga shrubland and sand-ridge hummock grassland. Nocturnal. Shelters beneath rocks and fallen timber; also uses spider burrows.

Tessellated Gecko ■ *Diplodactylus tessellatus* SVL 5.5cm

DESCRIPTION Body robust. Pale brown to dark grey, patternless to flecked. Pattern, if present, extends on to original tail. Regenerated tail same colour as body, with dark mottling. Underside white with distinctive grey blotches. Probably a species complex. **DISTRIBUTION** Eastern Australia, from near Horsham, VIC, north through most of

western NSW and QLD to Mt Isa, across Barkly Tableland, NT, and south to Port Augusta, SA. **HABITAT AND HABITS** Uses a wide variety of habitats, from dry woodland to mulga shrubland and black-soil plains, usually with heavy or clay-based soils. Nocturnal. Shelters beneath rocks and fallen timber. Also uses spider burrows and soil cracks. Usually seen on the ground at night hunting termites.

Eastern Stone Gecko ■ *Diplodactylus vittatus* SVL 5cm

DESCRIPTION Body robust. Dark brown to greyish-black. Pattern variable, consisting of broad pale blotches centred along spine, sometimes coalescing and forming a stripe. Pattern extends on to original tail. Regenerated tail same colour as body, with dark

mottling. Underside white. Probably a species complex. **DISTRIBUTION** Southern Australia, from Adelaide, SA, through central VIC, ACT and NSW, to Townsville, QLD. **HABITAT AND HABITS** Widespread, using heaths, grassland, mallee, brigalow, mulga and dry eucalypt forest. Nocturnal. Emerges from beneath rocks and timber to hunt invertebrates.

Reticulated Velvet Gecko ■ *Hesperoedura reticulata* SVL 7cm

DESCRIPTION Dorsally compressed. Light grey to black. Broad, dark-edged irregular
vertebral stripe. Pattern extends
on to original tail. Regenerated
tail same colour as body, with
dark mottling. Underside
white. Colour changes from
day to night. Darker by day
with patterning less prominent.
DISTRIBUTION Southern
WA, from Christmas Tree Well
to Zanthus. **HABITAT AND
HABITS** Restricted to heaths
and eucalypt forest. Nocturnal.
Often seen on trunks of smooth-
barked trees.

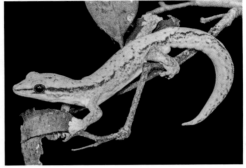

S Eipper

Gibber Gecko ■ *Lucasium byrnei* SVL 6cm

DESCRIPTION Body medium in build. Light reddish-brown to brown with pale yellow
blotches over body. Pattern extends on to original tail. Regenerated tail same colour as
body, with mottling. Underside white. **DISTRIBUTION** Avon Downs, NT, western QLD,
south over central and eastern SA, to Leeton, NSW. **HABITAT AND HABITS** Uses arid
habitats, from gibber desert and dry woodland, to mulga shrubland with heavy stony soils.
Nocturnal. Shelters beneath rocks and fallen timber. Usually seen on the ground at night
hunting invertebrates.

S Eipper

Beaded Gecko ■ *Lucasium damaeum* SVL 6cm

DESCRIPTION Body medium in build. Light reddish-brown to brown, with pale cream to yellow spots and flecks over body. Usually pale crown on head that extends into broken 'Y'-shaped stripe along spine. Pattern extends on to original tail. Regenerated tail same colour as body, with mottling. Underside white. Likely a species complex. **DISTRIBUTION** Much

of arid southern Australia, from Laverton, WA, through SA, into north-west VIC, western NSW and QLD, extending north to Epenarra, NT. **HABITAT AND HABITS** Uses a wide variety of habitats, from dry woodland and heaths, to mallee on sandy soils. Nocturnal. Shelters beneath rocks and fallen timber. Usually seen on the ground at night hunting.

Yellow-snouted Ground Gecko ■ *Lucasium occultum* SVL 6cm

DESCRIPTION Body medium in build. Dark purplish-brown, with pale cream to yellow spots and flecks over body. Pale blotches extend along spine. Pattern extends on to original tail. Regenerated tail same colour as body, with mottling. Tip of nose usually yellow. Underside white. **DISTRIBUTION** Top of the NT, between Litchfield and Kaplaga. **HABITAT AND HABITS** Found in open savannah woodland on cracking clay soils. Nocturnal. Shelters beneath fallen timber; usually seen on the ground at night hunting.

Box-patterned Gecko ■ *Lucasium steindachneri* SVL 6cm

DESCRIPTION Body medium in build. Pink, orange or reddish-brown to brown, with pale broad vertebral stripe that encloses 6–9 square-shaped blotches. Pattern extends on to original tail. Additional light spots and flecking sometimes present. Regenerated tail same colour as body, with mottling. Underside white. **DISTRIBUTION** Coen, through QLD and NSW, west of the GDR, to Balranald, NSW, and across into eastern SA. **HABITAT AND HABITS** Uses a wide variety of habitats, from dry woodland and brigalow, to mulga shrubland. Nocturnal. Shelters beneath rocks and fallen timber; usually seen on the ground at night hunting.

S Eipper

Crowned Gecko ■ *Lucasium stenodactylum* SVL 6cm

DESCRIPTION Body medium in build. Light reddish-brown to brown, with pale cream to yellow spots and flecks over body. Pale crown on head extends into an entire 'Y' shape on to nape, and becomes a vertebral stripe. Pattern extends on to original tail. Regenerated tail same colour as body, with mottling. Underside white. A species complex.

DISTRIBUTION Much of north and central Australia, from Shark Bay, WA, through SA, all the NT, western QLD, and western NSW to Bourke. **HABITAT AND HABITS** Uses a wide variety of habitats including dry woodland, heaths, sand-ridge desert and mallee on sandy soils. Nocturnal. Shelters beneath rocks and fallen timber; usually seen on the ground at night hunting.

A Elliott

Robust Velvet Gecko ■ *Nebulifera robusta* SVL 10cm

DESCRIPTION Body dorsally compressed. Light grey to charcoal. Series of large, pale grey, white to tan blotches roughly aligned along spine. Pattern extends on to original tail. Regenerated tail same colour as body, with dark mottling. Underside white. Massive

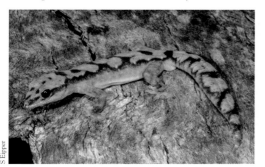

colour change from day to night. Darker by day, with patterning less prominent. **DISTRIBUTION** Eastern Australia, from Newcastle, NSW, to Capella, QLD. **HABITAT AND HABITS** Restricted to heaths and eucalypt forest. Nocturnal. Often seen on trunks of trees or rock faces, but some populations utilize houses. Feeds on invertebrates.

Beautiful Velvet Gecko ■ *Oedura bella* SVL 10cm

DESCRIPTION Body flattened. Grey to purplish-black with mix of vivid yellow and cream reticulations, spots and stripes covering body. Pattern extends on to original tail. Regenerated tail same colour as body, with similar pattern. Underside white. Juveniles banded yellow. **DISTRIBUTION** Northern Australia, from Sybella Creek, QLD, through the Gulf Country, to Borroloola, NT. **HABITAT AND HABITS** Restricted to rocky cliffs and outcrops and shale with tropical open woodland. Nocturnal. Often seen on rock faces. Feeds on invertebrates.

Inland Marbled Velvet Gecko ■ *Oedura cincta* SVL 10cm

DESCRIPTION Body flattened. Grey to purplish-black with mix of pale yellow and cream reticulations and spots that form broad bands covering body. Pattern extends on to original tail. Regenerated tail same colour as body, with similar pattern. Underside white. Juveniles banded yellow or white.

DISTRIBUTION Central Australia, from St George, QLD, west to Mt Doreen, NT, and south into the Flinders Range, SA. Also through western NSW. **HABITAT AND HABITS** Found in drier habitats, from mallee, heath and mulga, to open woodland and deserts. Nocturnal. Emerges from beneath tree bark and from rock crevices to hunt invertebrates.

Northern Spotted Velvet Gecko ■ *Oedura coggeri* SVL 10cm

DESCRIPTION Body flattened. Light grey to charcoal. Covered in small white and yellow flecks and occelli that sometimes coalesce into bands. Pattern extends on to original tail. Regenerated tail same colour as body, with dark mottling. Underside white. **DISTRIBUTION** North-east QLD, from Hopevale to Charters Towers. **HABITAT AND HABITS** Found in dry open woodland and eucalypt forest, usually with rock outcrops. Nocturnal. Often seen on rock faces, but some populations utilize trees.

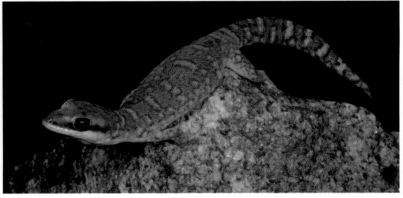

Fringe-toed Velvet Gecko *Oedura filicipoda* SVL 15cm

DESCRIPTION Body flattened. Light purplish-grey to charcoal. Covered in small white and yellow flecks and occelli that form broad bands. Pattern extends on to original tail

with more clearly defined banding. Regenerated tail same colour as body, with dark and light mottling. Underside white. **DISTRIBUTION** Northwest WA, from Mitchell Plateau to Derby. **HABITAT AND HABITS** Found in tropical open woodland with rock outcrops and gorges. Nocturnal. Often seen on rock faces and nearby vegetation.

Northern Marbled Velvet Gecko ■ *Oedura marmorata* SVL 10cm

DESCRIPTION Body flattened. Grey to purplish-black, with mix of vivid yellow and cream reticulations, spots and stripes covering body. Pattern extends on to original tail. Regenerated tail same colour as body, with similar pattern. Underside white. Juveniles banded yellow. Probably a species complex. **DISTRIBUTION** Northern Australia, from Borroloola, NT, to Kununurra, WA. **HABITAT AND HABITS** Found in tropical open woodland and rocky escarpments. Nocturnal. Emerges from beneath tree bark and from rock crevices to hunt invertebrates and other smaller geckos.

Elegant Velvet Gecko ■ *Oedura elegans* SVL 10cm

DESCRIPTION Body flattened. Light grey to charcoal. Series of large, dark-edged, white
to yellow blotches and streaks
on midline. Pattern extends on
to original tail. Regenerated
tail same colour as body, with
dark mottling. Underside white.
DISTRIBUTION Eastern
Australia, from Dubbo, NSW,
to Carnarvon Gorge, QLD.
HABITAT AND HABITS Found
in dry open woodland, heaths
and eucalypt forest, usually with
rock outcrops. Nocturnal. Often
seen on trees or rock faces, but
some populations utilize houses.

S Eipper

Southern Spotted Velvet Gecko ■ *Oedura tryoni* SVL 10cm

DESCRIPTION Body flattened. Light grey to charcoal. Covered in fine white and yellow
flecks and occelli that sometimes coalesce into streaks. Pattern extends on to original tail.
Regenerated tail same colour as body, with dark mottling. Underside white. A species
complex. **DISTRIBUTION** Eastern Australia, from Muswellbrook, NSW, to Blackdown
Tableland, QLD. **HABITAT AND HABITS** Found in dry open woodland, heaths and eucalypt
forest, usually with rock outcrops. Nocturnal. Population in the brigalow in Glenmorgan,
QLD, exclusively uses trees. Often seen on rock faces, but some populations utilize houses.

T Eipper

Giant Tree Gecko ■ *Pseudothedactylus australis* SVL 15cm

DESCRIPTION Body robust. Whitish-grey to tan or brown with fine black flecking, usually with light blotches. Edges of labials can be edged with black or brown. Regenerated

tail same colour as body. Terminal pad of lamellae beneath tail, used to assist in climbing. Underside white. Massive colour changes from day to night; much darker by day to creamish at night. **DISTRIBUTION** Coen to Prince of Wales Island in the Torres Strait. **HABITAT AND HABITS** Found in tropical open woodland and vine thickets. Nocturnal. Often seen at night on rock faces and nearby vegetation.

Northern Giant Cave Gecko ■ *Pseudothedactylus lindneri* SVL 17cm

DESCRIPTION Body robust. Light purplish-grey to charcoal. Covered in small white and yellow flecks. Series of purple to black blotches extend from nape to hips. These become bands on to original tail. Regenerated tail same colour as body. Terminal pad of lamellae beneath tail, used to assist in climbing. Underside white. Massive colour changes from day to night; much darker by day to yellow at night. **DISTRIBUTION** Arnhem Land and Kakadu NP, NT. **HABITAT AND HABITS** Found in tropical open woodland with rock outcrops and gorges. Nocturnal. Often seen at night on rock faces and nearby vegetation.

Brigalow Beaked Gecko ■ *Rynchoedura mentalis* SVL 5cm

DESCRIPTION Body medium in build. Light reddish-brown to brown with pale cream to yellow spots and flecks over body. Snout pointed, presumably to aid in catching prey.

Pattern extends on to original tail. Regenerated tail same colour as body, with mottling. Underside white. Eyes silver. **DISTRIBUTION** Arid QLD, from Longreach extending south to Quilpie. **HABITAT AND HABITS** Found in black-soil grassland, brigalow and mulga. Nocturnal. Shelters beneath rocks and fallen timber, and in spider burrows. Usually seen on the ground at night hunting termites.

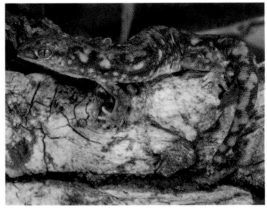

S Eipper

Eastern Beaked Gecko ■ *Rynchoedura ormsbyi* SVL 5cm

DESCRIPTION Body medium in build. Light reddish-brown to brown with pale cream to yellow spots and flecks over body. Snout pointed, presumably to aid in catching prey. Pattern extends on to original tail. Regenerated tail same colour as body, with mottling. Underside white. Possibly a species complex. **DISTRIBUTION** Much of arid eastern

Australia, from Mt Isa, QLD, extending south through western QLD and western NSW, to Hattah, VIC. **HABITAT AND HABITS** Uses a wide variety of habitats, including dry woodland, black-soil grassland, brigalow and mallee. Nocturnal. Shelters beneath rocks and fallen timber, and in spider burrows. Usually seen on the ground at night hunting termites.

S Eipper

Northern Spiny-tailed Gecko ■ *Strophurus ciliaris* SVL 10cm

DESCRIPTION Body robust. Highly variable, generally light grey to charcoal, brown to white. Usually mottled, becoming lighter on flanks. Long tubercles above eyes and on original tail, usually bright yellow, orange or black. Pattern extends on to original tail. Regenerated tail same colour as body, with dark mottling. Underside white. Colour changes from day to night; much darker by day, with patterning less prominent. Mouth orange. A species complex, currently with two described subspecies, *S. c. ciliaris* and *S. c.*

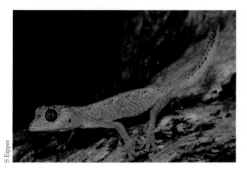

aberrans. **DISTRIBUTION** Arid and northern Australia, from Coolabah, NSW, through SA and WA, to Karratha; into northern Kimberley, across through the NT and western QLD, as far east as Hughenden. Isolated population on North West Cape, WA. **HABITAT AND HABITS** A generalist, found in dry open forest, heaths, deserts and tropical woodland. Nocturnal. Often seen hunting on the ground, rocks, trees and shrubs at night.

Jewelled Gecko ■ *Strophurus elderi* SVL 5cm

DESCRIPTION Body medium in build. Usually light grey to dark grey with fine white or yellow flecks edged with black. Pattern extends on to original tail. Regenerated tail same colour as body, without spotting. Underside lighter, with black flecking. Mouth colour blue. A species complex. Some animals are heavily spotted while others have fewer, much larger spots. **DISTRIBUTION** Southern Australia, from Cooma, NSW, north to Windorah, QLD, west through SA and southern NT, into WA between Fraser Range and

Edgar Ranges, east of Broome. **HABITAT AND HABITS** Found in sandy deserts, mallee and hummock grassland with spinifex. Nocturnal. Usually seen hunting in clumps of spinifex at night.

Southern Spiny-tailed Gecko ■ *Strophurus intermedius* SVL 7cm

DESCRIPTION Body medium in build. Variable; usually light grey to dark grey, sometimes with fine black reticulations and black flecks. Black markings usually form irregular zigzag pattern on back. Two rows of low tubercles on dorsolateral region of body that continue on to original tail. Pattern extends on to original tail. Regenerated tail same colour as body, with dark mottling or tubercles. Underside lighter with black flecking. Colour changes from day to night; much darker by day. Mouth blue. **DISTRIBUTION** Found across southern central Australia between Warialda, NSW to Wyperfield NP VIC, west through SA to Dundas NR WA and as far north as Alice Springs, NT. **HABITAT AND HABITS** A generalist, found in dry open forest, brigalow, mallee and mulga. Nocturnal. Often seen hunting on the ground, rocks, trees and shrubs at night.

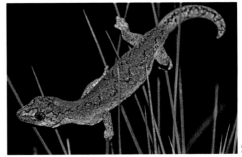

S Eipper

Kristin's Spiny-tailed Gecko ■ *Strophurus krisalys* SVL 7cm

DESCRIPTION Body medium in build. Highly variable; generally light grey to charcoal, brown to white. Usually mottled, becoming lighter on flanks. Long tubercles above eyes and on original tail usually bright yellow or black. Pattern extends on to original tail. Regenerated tail same colour as body, with dark mottling. Underside white. Colour changes from day to night; much darker by day with patterning less prominent. Mouth blue. Possibly a species complex with distinctive animals with much longer tubercles found in northern half of range. **DISTRIBUTION** QLD, from Lawn Hill NP, across to Croydon and south to Adavale. **HABITAT AND HABITS** A generalist, found in dry open forest, black-soil grassland, deserts and tropical woodland. Often seen hunting on the ground, rocks, trees and shrubs at night.

S Eipper

Northern Phasmid Gecko ■ *Strophurus taeniatus* SVL 6cm

DESCRIPTION Body light and elongated in build. Striped with light to dark grey, white and yellow; stripes can be edged with black. Pattern extends on to original tail. Regenerated

tail same colour as body, without striping. Underside lighter with black flecking. Mouth blue. **DISTRIBUTION** Northern Australia, from Longreach, QLD, through the NT, to King Leopold Ranges, WA. **HABITAT AND HABITS** Found in sandy deserts and hummock grassland with spinifex. Nocturnal. Often seen hunting in clumps of spinifex at night.

Golden-tailed Gecko ■ *Strophurus taenicauda* SVL 8cm

DESCRIPTION Body medium in build. Variable; usually white to dark grey with fine black reticulations and black flecks. Tail has bright orange to yellow, ragged stripe. Regenerated tail same colour as body. Underside lighter with black flecking. Colour changes from day to night; much darker by day. Mouth blue. Three subspecies currently recognized, separated by eye colour, tail markings and range. **DISTRIBUTION** Eastern Australia, from Moonie, through QLD west of the GDR, to Dingo. **HABITAT AND HABITS** A generalist found in dry open forest and brigalow. Nocturnal. Often seen hunting on the ground, rocks, trees and shrubs at night.

Eastern Spiny-tailed Gecko ■ *Strophurus williamsi* SVL 7cm

DESCRIPTION Body medium in build. Variable; usually light grey to dark grey, sometimes with fine black reticulations and black flecks. Two rows of low tubercles on dorsolateral region of body form four rows on to original tail. Pattern extends on to original tail. Regenerated tail same colour as body, with dark mottling or tubercles. Underside lighter with black flecking. Colour changes from day to night; much darker by day. Mouth blue. **DISTRIBUTION** Eastern Australia, from Townsville, through QLD west of the GDR, recorded at Hattah, VIC and across into eastern SA. **HABITAT AND HABITS** A generalist, found in dry open forest, brigalow, mallee and mulga. Nocturnal. Often seen hunting on the ground, rocks, trees and shrubs at night.

T Eipper

GEKKONID GECKOS

Marbled Gecko ■ *Christinus marmoratus* SVL 7cm

DESCRIPTION Body medium in build. White to dark grey with charcoal-coloured spots and streaks. Original tail has orange to yellow markings along midline. Regenerated tail same colour as body, without orange markings. Underside white. Colour changes from day to night; darker by day with patterning less prominent. Usually has prominent calcium deposits behind jaw. **DISTRIBUTION** Southern Australia, from Coonabarabran, NSW, south to Melbourne, VIC, and across to Lancelin, WA. **HABITAT AND HABITS** Found in dry open forest and gardens, usually on rocky outcrops. Nocturnal. Often seen under rocks by day, or hunting on rocks and trees at night.

S Eipper

Cooktown Ring-tailed Gecko ■ *Cyrtodactylus tuberculatus* SVL 17cm

DESCRIPTION Body robust. Alternating pattern of white and brown bands that are edged

in black. Upper body covered in tubercles. Tail-tip white. Toes have claws rather than pads. **DISTRIBUTION** North-east QLD, from Mt Leswell to Cape Melville. **HABITAT AND HABITS** Found in dry tropical woodland, rainforest and vine forest. Nocturnal. Emerges at night to sit upside down on trees, branches and rock faces near the ground, where it ambushes passing invertebrates and other geckos.

Northern Dtella ■ *Gehyra australis* SVL 10cm

DESCRIPTION Body robust. White to dark grey with charcoal-coloured spots and streaks. Underside white. Colour changes from day to night; darker by day with patterning more prominent. Occasionally completely white at night, without pattern. **DISTRIBUTION** North-west NT, from Timber Creek to Mataranka, and north to West Arnhem. **HABITAT AND HABITS** Occurs in tropical woodland, deserts and houses. Nocturnal. Often found under tree bark during the day, or hunting on trees at night.

Chain-backed Dtella ■ *Gehyra catenata* SVL 7cm

DESCRIPTION Body medium in build. White to dark grey with charcoal-coloured spots and streaks. Usually a dark, chain-like pattern along spine. Underside white. Colour changes from day to night; darker by day, with patterning more prominent. **DISTRIBUTION** Eastern Australia west of the GDR, QLD, between Charters Towers and Mitchell, and west to Idalia NP. **HABITAT AND HABITS** Found in dry open forest, brigalow and houses. Nocturnal. Seen under tree bark during the day, or hunting on trees at night.

Dubious Dtella ■ *Gehyra dubia* SVL 9cm

DESCRIPTION Body robust. White to dark grey with charcoal-coloured spots and streaks. Underside white. Colour changes from day to night; darker by day with patterning more prominent. Probably a species complex. **DISTRIBUTION** Eastern Australia, from central NSW between Yathong NR and Quirindi, through QLD, to the Gulf of Carpentaria, Cape York and Torres Strait Islands. Also PNG. **HABITAT AND HABITS** Found in forest, mallee, brigalow, mulga, deserts and houses. Nocturnal. Often seen under rocks or tree bark during the day, or hunting on rocks and trees at night.

Northern Spotted Dtella ■ *Gehyra nana* SVL 6cm

DESCRIPTION Body medium in build. White, yellow to pale orange with dark grey to purple spots and streaks. Underside white. Colour changes from day to night; darker by day with patterning more prominent. **DISTRIBUTION** North-west QLD, from Riversleigh, QLD, across the Top End region of the NT, to Kimbolton in the Kimberley region of WA. **HABITAT AND HABITS** Found in tropical open woodland, rock escarpments, deserts and houses. Nocturnal. Often seen under rocks during the day, or hunting on rock faces at night.

Robust Dtella ■ *Gehyra robusta* SVL 9cm

DESCRIPTION Body robust. White, yellow to pale orange with dark grey to purple spots and streaks. Underside white. Colour changes from day to night; darker by day with patterning more prominent. Probably a species complex. **DISTRIBUTION** North-west QLD, from Cawnpore Lookout to Dajarra, through western QLD, to Wollogorang, NT. **HABITAT AND HABITS** Found in tropical open woodland, rock escarpments, deserts and houses. Nocturnal. Often seen under rocks during the day, or hunting on rock faces at night.

Variegated Dtella ■ *Gehyra variegata* SVL 7cm

DESCRIPTION Body medium in build. White to dark grey with charcoal-coloured spots and streaks. Underside white.
Colour changes from day to night; darker by day with patterning more prominent. **DISTRIBUTION** Southern half of WA, from Karratha to the Darling Range east of Perth, across to Yalata, SA. **HABITAT AND HABITS** Found in dry open forest, mallee, mulga, deserts and houses. Nocturnal. Often seen under rocks or tree bark during the day, or hunting on rocks and trees at night.

S Eipper

Variable Dtella ■ *Gehyra versicolor* SVL 7cm

DESCRIPTION Body medium in build. White to dark grey with charcoal-coloured spots and streaks. Underside white. Colour changes from day to night; darker by day with patterning more prominent. **DISTRIBUTION** Eastern and central Australia west of the
GDR, from Mt Isa, QLD, south to Hattah, VIC, and west across to Bates, SA; through central SA and into the NT, to the Macdonnell Ranges and north-east to Barkly Tableland. **HABITAT AND HABITS** Found in dry open forest, mallee, brigalow, mulga, deserts and houses. Nocturnal. Often seen under rocks or tree bark during the day, or hunting on rocks and trees at night.

S Eipper

Asian House Gecko ■ *Hemidactylus frenatus* SVL 9cm

DESCRIPTION Body medium in build. White to dark grey with dark brown spots and streaks. Underside white. Colour changes from day to night; darker by day with patterning less prominent at night. Almost white at night. Low tubercles on original tail only.
DISTRIBUTION Urban areas north of Coffs Harbour, NSW, through QLD, NT and WA; also Christmas and Norfolk Islands. Worldwide in tropical and subtropical cities.
HABITAT AND HABITS Found in forest, brigalow, mulga, deserts and houses. Often seen on man-made structures. Invasive species first recorded in Australia in the 1970s in Darwin. Nocturnal. Utters a loud *chuck-chuck-chuck* during the day, usually from cover, to maintain its home range. Feeds on invertebrates. Lays two eggs per clutch with up to five clutches per season.

T Eipper

Fox Gecko ■ *Hemidactylus garnotii* SVL 9cm

DESCRIPTION Body medium in build. White to dark grey with dark brown spots and streaks. Underside yellow. Colour changes from day to night; darker by day with patterning less prominent at night. Almost white or yellowish at night. Low tubercles on original tail only.
DISTRIBUTION Urban areas near Manly in Sydney, NSW.
HABITAT AND HABITS Found in forests and on buildings. Nocturnal. Often seen on man-made structures. Invasive species first recorded in Australia in 2017 at Manly, Sydney. Feeds on invertebrates. Lays two eggs per clutch.

S Eipper

Bynoe's Gecko ▪ *Heteronotia binoei* SVL 7cm

DESCRIPTION Body medium in build. Covered in low tubercles and can be flecked, banded and blotched, or without pattern. Underside white. A large species complex. **DISTRIBUTION** Across mainland Australia west of the GDR, and excluding cool south-east of NSW and VIC. Also PNG. **HABITAT AND HABITS** Found in dry open forest, cracking clay plains, brigalow, heath, mallee, mulga, deserts and houses. Nocturnal. Often seen under rocks or tree bark during the day, or hunting on rocks at night. Some populations are parthenogenetic.

Pale-headed Gecko ▪ *Heteronotia fasciolata* SVL 8cm

DESCRIPTION Body medium in build. Covered in low tubercles, banded in shades of brown, cream and white. Underside white. **DISTRIBUTION** Around Alice Springs and the West Macdonnell Ranges in the NT; disjunct population at Mt Isa, QLD. **HABITAT AND HABITS** Found in dry open forest, mallee, mulga, deserts and houses. Nocturnal. Often seen under rocks or tree bark during the day, or hunting on rocks at night.

Mourning Gecko ■ *Lepidodactylus lugubris* SVL 7cm

DESCRIPTION Body medium in build and flattened. White to dark grey with dark brown blotches edged with black centred along midline. Underside white. Colour changes from day to night; much darker by day with stronger patterning. Almost white at night. **DISTRIBUTION** Urban areas in QLD north of Townsville. Recently seen in Darwin, NT, Heron Island and on the Sunshine Coast, QLD. **HABITAT AND HABITS** Found in forests and on buildings. Nocturnal. Possibly an invasive species.

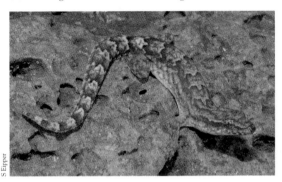

Southern Cape York Nactus ■ *Nactus cheverti* SVL 8cm

DESCRIPTION Body medium in build. Olive-green to grey-brown, usually flecked with dark pigment and irregular bands centred along midline. Tail banded with black. Top of head often pale yellow. Iris reddish. Covered in low tubercules that form rows. Underside white to grey. **DISTRIBUTION** North-east QLD, from Innisfail to near Coen. **HABITAT AND HABITS** Found in tropical woodland, rainforest and vine thickets. Nocturnal. Often seen under rocks or logs during the day, or hunting on the ground at night.

48

Black Mountain Gecko ■ *Nactus galgajuda* SVL 7cm

DESCRIPTION Body medium in build. Charcoal-grey to black, usually flecked with white pigment and irregular white bands centred along midline. Tail banded with white. Iris reddish. Covered in low tubercules that form rows. Underside white to grey. **DISTRIBUTION** North-east QLD at Kalkajaka NP, near Rossville. **HABITAT AND HABITS** Found on black granite boulders and surrounding tropical woodland and vine thickets. Nocturnal. Usually only seen hunting on boulders at night. Probably shelters in rock crevices by day.

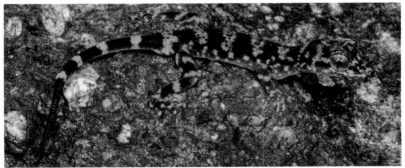

S Eipper

FLAT-FOOTED LIZARDS

Red-tailed Worm Lizard ■ *Aprasia inaurita* SVL 14cm

DESCRIPTION Top of body cream to beige on front half; rear half orange to pink. Underside like upper body. **DISTRIBUTION** Between Lake Hindmarsh, VIC, north to Balranald, NSW, across southern SA to Caiguna, WA. **HABITAT AND HABITS** Lives beneath mallee roots, fallen timber and leaf litter in mallee woodland with a spinifex understorey. Nocturnal, but also basks beneath cover in leaf litter. Not much is known about this secretive species – it has been noted in the diet of the Mallee Black-headed Snake *Suta nigriceps*. Feeds on ants.

N Gale

Pink-tailed Worm Lizard ■ *Aprasia parapulchella* SVL 14 cm

DESCRIPTION Top of body beige to dark brown with fine black specks arranged in rows that can form fine broken stripes. Tail from above has pink flush over the brown.

Underneath cream to white, with tail-tip underside bright pink. **DISTRIBUTION** Between Bendigo, VIC, and Goulburn, NSW, including the ACT. **HABITAT AND HABITS** Lives beneath stones and in ant nests in dry eucalypt forest, usually on rocky hillsides. Nocturnal, but also basks by day. Feeds on ants.

Lined Worm Lizard ■ *Aprasia striolata* SVL 14cm

DESCRIPTION Top of body yellowish-cream to beige, with or without 3–5 black longitudinal stripes. Some populations in SA and WA completely devoid of stripes.

Underside similar to upper body. **DISTRIBUTION** Two populations. Eastern population between Penthurst, VIC, across to Ceduna, SA; Western population in WA, between Yellanup and Balladonia. **HABITAT AND HABITS** Lives in mallee woodland, grassland and heaths beneath mallee roots, fallen timber and leaf litter. Nocturnal, but also basks beneath cover in leaf litter. Feeds on ants.

Marble-faced Delma ■ *Delma australis* SVL 16cm

DESCRIPTION Head and forebody white with black flecking. Body grey to reddish-brown. Underside light grey to yellowish; speckled with black beneath head. Potentially a species complex. **DISTRIBUTION** Five populations across southern Australia. Western population from Wooramel to Cocklebiddy, south-west WA; central population from Marlinga, SA, to Parachilna, SA, and north to Alice Springs, NT; eastern population between Adelaide, SA, and Swan Hill, VIC. Isolated populations near Round Hill, NSW, and Shot Hole Canyon, WA. **HABITAT AND HABITS** Lives in mallee woodland, grassland and heath beneath mallee roots and fallen timber, inside stick-ant nests and beneath leaf litter. Nocturnal, but also basks beneath cover in leaf litter. Feeds on invertebrates.

S Eipper

Spinifex Delma ■ *Delma butleri* SVL 16cm

DESCRIPTION Body usually olive or greyish-brown. Sometimes has white to cream bars on lips and on to neck. Usually white ring around each eye. Underside light grey to yellowish. Potentially a species complex, with consistent differences between eastern and western populations. **DISTRIBUTION** Southern Australia, from Exmouth, WA, through central Australia, to Ouyen, VIC. **HABITAT AND HABITS** Lives in mallee woodland, grassland and heaths beneath mallee roots and fallen timber, inside stick-ant nests and beneath leaf litter. Nocturnal, but also basks perched in clumps of spinifex. Feeds on invertebrates.

S Eipper

Fraser's Delma ■ *Delma fraser* SVL 18cm

DESCRIPTION Body usually greyish-brown. Sometimes black bars on lips and on to neck. Top of head usually dark brown to black in juveniles. Underside light grey to yellowish.

DISTRIBUTION Southwestern WA, from Shark Bay to Balladonia, WA. **HABITAT AND HABITS** Lives in woodland, rocky hillsides and heaths, beneath rocks and fallen timber, and inside stick-ant nests. Diurnal. Secretive in nature, usually basking from within clumps of grass or in vegetation on edge of shelter. Feeds on invertebrates.

Striped Delma ■ *Delma impar* SVL 16cm

DESCRIPTION Generally olive or greyish-brown, usually with prominent stripes running along body. Head can be yellow. Usually a white ring around each eye. Underside light whitish-grey to yellowish. **DISTRIBUTION** Southeastern Australia, from Melbourne, VIC, through ACT and NSW, to Cooma. Isolated population at Muswellbrook, NSW, may represent an undescribed taxon. **HABITAT AND HABITS** Lives in grassland and heaths beneath rocks, fallen timber and leaf litter. Diurnal. Secretive in nature, usually basking from within clumps of grass or in vegetation on the edge of shelter. Feeds on invertebrates.

Olive Delma ■ *Delma inornata* SVL 16cm

DESCRIPTION Body generally olive or greyish-brown. Skin between scales black. Usually a white ring around each eye. Underside light grey to yellowish. Potentially a species complex, with consistent differences between northern and southern populations. **DISTRIBUTION** Southeastern Australia, from Ballarat, VIC, through NSW and ACT, to Acland, QLD. **HABITAT AND HABITS** Lives in grassland and heaths beneath rocks, fallen timber and leaf litter. Diurnal. Secretive in nature, usually basking from within clumps of grass or in vegetation on the edge of shelter. Feeds on invertebrates.

S Eipper

Atherton Delma ■ *Delma mitella* SVL 29cm

DESCRIPTION Largest of all the delmas. Body generally olive or greyish-brown. Skin between scales black. Usually a white ring around each eye. Underside light green to yellowish. **DISTRIBUTION** North-east QLD, from Paluma to Mareeba. **HABITAT AND HABITS** Lives in tropical woodland and tall eucalyptus forest with dense understorey. Diurnal. Secretive in nature, and usually seen crossing paths or moving on the forest floor. Has been found using man-made cover. Feeds on invertebrates.

A Zimmy

Sharp-snouted Delma ▪ *Delma nasuta* SVL 20cm

DESCRIPTION Body generally olive or greyish-brown. Skin between scales black. Usually a white ring around each eye. Head elongated. Underside light grey to yellowish. Tail can be very long, up to three times body length. **DISTRIBUTION** Arid Australia, from Shark Bay south to Cundeelee, WA, across through northern SA to Stonehenge, QLD, north to Limmen, NT. **HABITAT AND HABITS** Lives in tropical woodland, rocky outcrops and sandy deserts with spinifex. Cathemeral. Secretive in nature, usually basking from within clumps of vegetation or on edge of shelter. Commonly seen crossing roads on warm nights. Feeds on invertebrates.

Leaden Delma ▪ *Delma plebeia* SVL 17cm

DESCRIPTION Body generally greyish-brown. Sometimes has black bars on lips and on to neck. Southern animals can have orange-coloured throats. Top of head usually

dark brown to black in juveniles. Underside light grey to yellowish. **DISTRIBUTION** Eastern Australia, from Singleton, NSW, to Charleville, QLD. **HABITAT AND HABITS** Lives in woodland, grassland and brigalow beneath rocks and fallen timber. Diurnal. Secretive in nature, usually basking from within clumps of grass or in vegetation on edge of shelter. Feeds on invertebrates.

Black-necked Delma ■ *Delma tincta* SVL 19cm
(Excitable Delma)

DESCRIPTION Body variable, from sandy-yellow to reddish-brown, to grey. Top of head usually dark brown to black in juveniles, with dark bar across nape. Underside light grey to yellowish. **DISTRIBUTION** Northern and central Australia north of Fraser Island, QLD, to Tamworth, NSW, through SA, the NT, and WA north of Paynes Find. **HABITAT AND HABITS** Lives in woodland, grassland and deserts beneath rocks and fallen timber. Cathemeral. Secretive in nature, usually basking from within clumps of grass or in vegetation on edge of shelter. Also known as the Excitable Delma, due to its habit of leaping at would-be attackers. Feeds on invertebrates.

S Eipper

Collared Delma ■ *Delma torquata* SVL 14cm

DESCRIPTION The smallest species of *Delma*, often mistaken for hatchling brown snake. Body usually greyish-brown. Sometimes has black bars on lips and on to neck. Top of head dark brown to black, with two dark bars across nape edged with orange. Underside light grey to yellowish; underneath of head speckled with black. **DISTRIBUTION** South-east QLD, from Brisbane to Calliope. **HABITAT AND HABITS** Lives in woodland, grassland and brigalow beneath rocks and fallen timber. Diurnal. Secretive in nature, usually basking from within clumps of grass or in vegetation on edge of shelter. Feeds on invertebrates.

S Eipper

Burton's Legless Lizard ■ *Lialis burtonis* SVL 33cm

DESCRIPTION Very variable, ranging from white to black, and every shade of yellow, orange-red and brown, with or without pattern of spots or stripes. Tail-tip can be brightly

coloured. Underside just as variable as dorsum. Head elongated, with very pointed snout, which is hinged to clamp down on prey. **DISTRIBUTION** Most of mainland Australia, except far south-east SA, southern VIC and far south-east NSW. **HABITAT AND HABITS** Lives in woodland, grassland and deserts beneath rocks and fallen timber. Cathemeral. Secretive in nature. Often mistaken for a snake. Uses both active pursuit and ambushing as strategies for prey capture. Feeds on other lizards and occasionally small snakes.

Above and below, left and right: colour variations

Bronzeback ■ *Ophidiocephalus taeniatus* SVL 10cm

DESCRIPTION Usually pale yellow, fawn to bronze above, with reddish-brown to dark brown sides. Head white to grey and pointed to aid in burrowing. Underside greyish

with light-coloured flecking that extends on to lower flanks. **DISTRIBUTION** Central Australia, between Charlotte Waters, NT, south to Coober Pedy, SA. **HABITAT AND HABITS** Lives in mulga-dominated woodland and dry creek beds. Nocturnal. Fossorial. Found beneath and within deep leaf litter mats that form beneath low trees in its range. Feeds on small invertebrates.

Brigalow Scaly-foot ■ *Paradelma orientalis* SVL 21cm

DESCRIPTION Body variable, from dark brown to grey. Head usually dark brown to black in juveniles, with yellow crown. Underside white to yellowish or grey, with or without dark flecking. **DISTRIBUTION** Most of eastern QLD, from Westmar, north to Mt Larcom and west to Bollon. **HABITAT AND HABITS** Lives in woodland, grassland and brigalow beneath rocks, fallen timber and leaf litter. Nocturnal. Feeds on invertebrates, especially spiders; also licks soft fruits and nectar. Noted to climb into trees to eat *Acacia* sap.

S Eipper

Keeled Legless Lizard ■ *Pletholax gracilis* SVL 10cm

DESCRIPTION Olive to greyish-brown, faintly striped with grey. Posterior third sometimes yellowish, particularly in juveniles. Head strongly pointed; dark-coloured above and bright yellow below that extends along lower forebody on to flanks. Underside greyish with light-coloured flecking that extends on to lower flanks. Scales, including ventrals, strongly keeled. **DISTRIBUTION** Along Western Australian coast, between Eneabba and Mandurah. **HABITAT AND HABITS** Lives in sandy heaths and dry coastal woodland. Diurnal. Found active by day on the surface and moving through vegetation; also moving below the soil surface and in rotten logs. Thought to feed on termites.

A McNab

Common Scalyfoot ■ *Pygopus lepidopodus* SVL 35cm

DESCRIPTION Usually grey or shade of reddish-brown to beige. Some individuals reddish-brown with grey heads; others plain, and some heavily patterned with purplish to black spots edged with pink. Underside white to yellowish or grey, with or without pattern of dark flecking. **DISTRIBUTION** Possibly now two populations that were historically continuous in central VIC. One across most of southern Australia, from Shark Bay, WA, to Bendigo, VIC; eastern

population from Melbourne, VIC, to Yeppoon, QLD. **HABITAT AND HABITS** Lives in woodland, grassland, heaths and deserts beneath rocks, fallen timber and leaf litter. Cathemeral. Secretive in nature, and usually seen while crossing roads at night. Commonly mistaken for a small snake. Feeds on invertebrates, especially spiders; also licks soft fruits and nectar.

Western Hooded Scalyfoot ■ *Pygopus nigriceps* SVL 20cm

DESCRIPTION Usually sandy-yellow to reddish-brown to grey. Top of head generally dark brown to black in juveniles, with dark bar across nape. Some individuals are plain while others are heavily patterned. Underside white to yellowish or grey, with or without pattern of dark flecking. **DISTRIBUTION** Most of northern and central Australia, from Quilpie, QLD, west to Shark Bay, WA, extending as far north as Katherine, NT, and as far south

as the Gawler Ranges, SA. **HABITAT AND HABITS** Lives in tropical woodland, grassland, heaths and deserts beneath rocks, fallen timber and leaf litter. Cathemeral. Secretive in nature, and usually seen while crossing roads at night. Commonly mistaken for a small snake. Feeds on invertebrates, especially spiders; also licks soft fruits and nectar.

Eastern Hooded Scalyfoot ■ *Pygopus schraderi* SVL 25cm

DESCRIPTION Body variable, from sandy-yellow to reddish-brown, to grey. Top of head usually dark brown to black in juveniles, with dark bar across nape. Some individuals are plain while others are heavily patterned. Underside white to yellowish or grey, with or without pattern of dark flecking. **DISTRIBUTION** Most of eastern Australia, from Terrick Terrick NP, VIC, north through NSW and QLD, to Karumba and west to Tablelands, NT, in north, and as far west as Coober Pedy, SA, in south. **HABITAT AND HABITS** Lives in woodland, grassland, heaths, mallee, brigalow and deserts beneath rocks, fallen timber and leaf litter. Cathemeral. Secretive in nature, and usually seen while crossing roads at night. Commonly mistaken for a small snake. Feeds on invertebrates, especially spiders; also licks soft fruits and nectar.

S Eipper

SKINKS

Eastern Three-lined Skink ■ *Acritoscincus duperreyii* SVL 8.5cm

DESCRIPTION Body robust. Dark brown to greyish-silver. Prominent black vertebral stripe. Pale-edged dark dorsolateral stripe that extends broadly down flanks. Usually a pale low lateral stripe extends from sides of head to tail. Throat of breeding males red to orange. Cream to white underside. Four limbs with five digits on each foot. **DISTRIBUTION** Eastern SA, through southern VIC, north and east TAS, and high country of southern NSW north to Newnes. **HABITAT AND HABITS** A true generalist, using wet forests, open woodland and urban environments. Diurnal. Shelters beneath logs and rocks, and often seen beneath pots in gardens. Eats invertebrates. Lays 3–9 eggs per clutch.

A Elliott

Red-throated Skink ■ *Acritoscincus platynota* SVL 8cm

DESCRIPTION Body robust. Dark brown to greyish-silver. Pale-edged dark dorsolateral stripe that extends broadly down sides. Both sexes have reddish-orange throat that is more intense in breeding males. Cream to white underside. Four limbs with five digits on

each foot. **DISTRIBUTION** Eastern VIC, up through eastern NSW, into granite outcrops in south-east QLD, near Stanthorpe. **HABITAT AND HABITS** A true generalist, using wet forests, open woodland and urban environments. Diurnal. Shelters beneath logs and rocks, and often seen beneath pots in gardens. Eats invertebrates. Lays 3–11 eggs in a clutch, often communally.

McCoy's Skink ■ *Anepischetosia maccoyi* SVL 5cm

DESCRIPTION Body elongated. Dark brown to brown, usually with grey flecking that becomes more prominent towards tail. Lower flanks and underside yellow. Four short limbs with five digits on each foot. **DISTRIBUTION** Southeastern NSW and VIC from the Illawarra region to Millicent, SA. **HABITAT AND HABITS** Found in wet forests, closed woodland and moist urban environments. Diurnal. Shelters beneath logs and rocks. Eats invertebrates. Lays 1–5 eggs per clutch.

Swanson's Worm Skink ■ *Anomalopus swansoni* SVL 10cm

DESCRIPTION Body elongated. Pinkish-brown to brown that becomes greyish-black on tail-tip. Completely limbless.
DISTRIBUTION Hunter Valley from Newcastle to Quirindi, NSW.
HABITAT AND HABITS Found in open woodland and rocky slopes. Nocturnal. Fossorial. Shelters beneath logs and rocks. Can be seen on the surface at night, particularly after rain. Commonly mistaken for a small snake. Eats invertebrates. The only member of the genus that is a livebearer. Gives birth to 2–3 live young per litter.

S Eipper

Verreaux's Skink ■ *Anomalopus verreauxii* SVL 12cm

DESCRIPTION Body elongated. Dark brown to black. Lower flanks and underside yellow. Yellow band across nape. Young animals and juveniles brighter than adults. Four rudimentary limbs that are effectively useless. **DISTRIBUTION** Eastern QLD and NSW, between Proserpine and Walcha. **HABITAT AND HABITS** A true generalist, using wet forests, open woodland and urban environments. Nocturnal. Fossorial. Shelters beneath logs and rocks, and often seen beneath pots in gardens. Often seen by home owners after they come into a house during the night, or when disturbed while gardening in south-east QLD. Commonly mistaken for a small snake. Eats invertebrates. Lays 1–6 eggs per clutch.

T Eipper

Major Skink ■ *Bellatorias frerei* SVL 30cm

DESCRIPTION Body robust. Tan to dark brown; flanks often have broad reddish-brown stripe with white spotting that is more prominent in juveniles. Some individuals have up to eight thin black stripes running parallel to spine. Underside pink to yellowish.

Four limbs with five digits on each foot. Potentially a species complex with distinct forms. **DISTRIBUTION** Northeastern NSW and QLD, between Bellingen and Sabai Island in the Torres Strait. Also southern PNG. **HABITAT AND HABITS** Restricted to rainforest, vine scrub and closed forest. Diurnal. Lives in family groups, usually in burrows by day. Omnivorous, eating insects, fungi and fruits. Gives birth to 3–10 live young per litter.

Land Mullet ■ *Bellatorias major* SVL 39cm

DESCRIPTION Body robust. Dark brown to black. Young animals and juveniles spotted with white. Juveniles can have reddish tint on lower flanks. Underside pink to yellowish. Four limbs with five digits on each foot. Australia's largest skink. **DISTRIBUTION**

Eastern NSW and QLD, between Kenilworth and Ourimbah. **HABITAT AND HABITS** Restricted to rainforest and wet eucalypt forest. Diurnal. Lives in family groups, usually in burrows formed under large fallen trees and among tree buttresses. Often seen by bushwalkers while basking by day. Omnivorous, eating insects, fungi and fruits. Gives birth to 4–11 live young per litter.

Garden Calyptotis ■ *Calyptotis scutirostrum* SVL 6cm

DESCRIPTION Body medium in build. Pale brown to grey with fine black flecking. Usually has black, ragged edged dorsolateral stripe. Flanks dark brown to charcoal with yellowish to cream spots. Underside of tail reddish to bright orange. Four short limbs with five digits on each foot. **DISTRIBUTION** Eastern NSW, from Dorrigo to Gympie, QLD. **HABITAT AND HABITS** Found in rainforest, and closed and open woodland, as well as urban environments. Diurnal. Shelters beneath logs and rocks. Eats invertebrates. Lays 2–4 eggs per clutch.

S Eipper

Coventry's Skink ■ *Carinascincus coventryi* SVL 5cm

DESCRIPTION Body robust. Dark brown to brown. Usually has grey flecking that becomes more prominent towards tail. Bronze-edged dorsolateral stripe above dark brown flanks. Underside white. Four limbs with five digits on each foot. **DISTRIBUTION** Along the GDR in southeastern NSW, from the Jenolan Caves to the Grampians, VIC. Isolated population near the Otways, VIC. **HABITAT AND HABITS** Found in rainforest, wet forests, closed woodland and moist urban environments. Diurnal. Shelters beneath logs and rocks. Usually most commonly seen while basking in forest clearings. Eats invertebrates. Gives birth to 2–7 live young per litter.

S Scott

Metallic Skink ■ *Carinascincus metallicus* SVL 8.5cm

DESCRIPTION Body robust. Dark brown to greyish-silver. Sometimes has prominent black vertebral stripe. Pale-edged dark dorsolateral stripe that extends broadly down sides. Usually a pale low lateral stripe extending from sides of head to tail. Throat of breeding

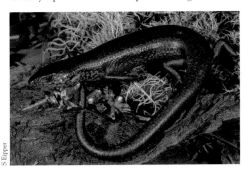

males red to orange. Cream to orange underside. Four limbs with five digits on each foot. **DISTRIBUTION** Over all TAS and neighbouring southern VIC. **HABITAT AND HABITS** A true generalist, using rainforest, wet forests, open woodland, grassland and urban environments. Diurnal. Shelters beneath logs and rocks, and often seen beneath pots in gardens. Eats invertebrates. Gives birth to 2–7 live young per litter.

S Eipper

Ocellated Skink ■ *Carinascincus ocellatus* SVL 8.5cm

DESCRIPTION Body robust. Dark brown to greyish-tan. Prominent cream to white spots over upper body. Pattern brighter on juveniles than adults. Cream to white underside. Four limbs with five digits on each foot. **DISTRIBUTION** Eastern Bass Strait Islands and all TAS bar south-west corner. **HABITAT AND HABITS** Lives in grassland, using wet forests, open woodland and urban environments. Diurnal. Shelters predominantly under rocks but also utilizes fallen timber and tree bark. Also uses crevices in rocks and logs. Eats invertebrates. Gives birth to 2–7 live young per litter.

A McNab

Two-spined Rainbow Skink ■ *Carlia amax* SVL 4cm

DESCRIPTION Body medium in build. Head generally copper coloured. Body silver to greyish-brown with black, grey and white flecking. Usually a pale streak from rostral scale under eye to ear. Underside white. Two keels on each scale. Four limbs with four digits on front feet and five on rear feet. **DISTRIBUTION** From the Kimberleys, WA, across northern Australia, to Mt Isa, QLD. **HABITAT AND HABITS** Found in rocky gorges and tropical open woodland on stony soils. Diurnal. Often seen waving tail while foraging, presumably to communicate. Shelters beneath logs and rocks. Eats invertebrates. Lays 1–6 eggs per clutch.

S Eipper

Lined Rainbow Skink ■ *Carlia jarnoldae* SVL 5cm

DESCRIPTION Body medium in build. Usually coppery-brown. Dark streak with pale lower edge from rostral scale along sides and on to lateral parts of tail. Breeding males have 5–7 black stripes along back; upper flanks dotted in fine blue to green spots; flanks behind armpits orange-red. Underside white. Three keels on each scale. Four limbs with four digits on front feet and five on rear feet. **DISTRIBUTION** North-east QLD, between Heathlands and Collinsville. **HABITAT AND HABITS** Found in open woodland on stony soils and in forests bordering rainforest. Diurnal. Often seen waving tail while foraging, presumably to communicate. Shelters beneath logs and rocks. Eats invertebrates. Lays 2–4 eggs per clutch.

S Eipper

Striped Rainbow Skink ■ *Carlia munda* SVL 4.5cm

DESCRIPTION Body medium in build. Silver to greyish-brown with black, grey and white flecking. Prominent pale streak from rostral scale under eye to midway between hip and shoulder. Breeding males have orange to red flanks with white flecking. Underside white. No keels on each scale. Four limbs with four digits on front feet and five on rear feet.

DISTRIBUTION From the Kimberleys, WA, across northern and eastern Australia, to

Ipswich, QLD. Isolated population between the Pilbara and North West Cape, WA, may represent an undescribed taxon. **HABITAT AND HABITS** Found in rocky gorges, spinifex grassland and open woodland on stony soils. Diurnal. Often seen waving tail while foraging, presumably to communicate. Shelters beneath logs and rocks. Eats invertebrates. Lays 2–5 eggs per clutch.

Open-litter Rainbow Skink ■ *Carlia pectoralis* SVL 5cm

DESCRIPTION Body medium in build. Silver to greyish-brown with black, grey and white flecking. Breeding males have 2–3 orange stripes on flanks. Underside white. Three keels

on each scale. Four limbs with four digits on front feet and five on rear feet. **DISTRIBUTION** QLD, from Brisbane, north to Dingo and west to St George. **HABITAT AND HABITS** Found in heaths, brigalow, grassland and open woodland on stony soils. Diurnal. Shelters beneath logs and rocks. Eats invertebrates. Lays 2–5 eggs per clutch.

Blue-throated Rainbow Skink ■ *Carlia rhomboidalis* SVL 6cm

DESCRIPTION Body medium in build. Dark brown to brown. Copper-coloured head. Usually has grey/black flecking that forms poorly defined pair of paravertebral stripes. Bronze-edged dorsolateral stripe. Flanks brown. Underside white. Throat and lower part of face bright blue. Bright red from ears to shoulder. Four limbs with four digits on front feet and five on rear feet. **DISTRIBUTION** Northeastern QLD, between Magnetic Island and Sarina. **HABITAT AND HABITS** Found in rainforests and neighbouring woodland. Diurnal. Shelters beneath logs and rocks. Most commonly seen while basking on rainforest trails. Eats invertebrates. Lays 2–5 eggs per clutch.

S Eipper

Northern Red-throated Skink ■ *Carlia rubrigularis* SVL 6cm

DESCRIPTION Body medium in build. Dark brown to brown. Copper-coloured head. Generally has grey to black flecking. Bronze-edged dorsolateral stripe. Usually has darker flanks. Underside white with pink-coloured throat that is bright red in breeding males. Four limbs with four digits on front feet and five on rear feet. **DISTRIBUTION** Northeastern QLD, between Townsville and Malanda. **HABITAT AND HABITS** Found in rainforests and neighbouring woodland. Diurnal. Shelters beneath logs and rocks. Most commonly seen while basking on rainforest trails. Eats invertebrates. Lays 2–5 eggs per clutch.

S Eipper

Southern Rainbow Skink ■ *Carlia tetradactyla* SVL 7cm

DESCRIPTION Body medium in build. Silver to greyish-brown with pair of black paravertebral stripes running down back. Breeding males have two orange stripes on flanks and teal throat, flushing on to forelimbs. Underside white. Three keels on each scale. Four limbs with four digits on front feet and five on rear feet. **DISTRIBUTION** Along

western slopes of the GDR and neighbouring plains, from near Benalla, VIC, through NSW, to Acland, QLD. **HABITAT AND HABITS** Found in grassland, open woodland and urban environments. Diurnal. Shelters beneath logs and rocks, as well as man-made debris. Eats invertebrates. Lays 2–7 eggs per clutch.

Tussock Rainbow Skink ■ *Carlia vivax* SVL 5cm

DESCRIPTION Body medium in build. Silver to greyish-brown with black, grey and white flecking. Breeding males have 2–3 orange stripes on flanks. Underside white. Three keels on each scale. Four limbs with four digits on front feet and five on rear feet.

DISTRIBUTION Along the GDR and neighbouring slopes in QLD and NSW, from Muswellbrook north to Prince of Wales Island in the Torres Strait, QLD. **HABITAT AND HABITS** Found in heath, brigalow, grassland, open woodland and urban environments. Diurnal. Shelters beneath logs and rocks, as well as man-made debris. Eats invertebrates. Lays 2–5 eggs per clutch.

Limbless Snake-tooth Skink ■ *Coeranoscincus frontalis* SVL 30cm

DESCRIPTION Body elongated. Dark brown, greyish to purplish-black, becoming lighter towards tail. Usually pale bar across snout. Lower flanks and underside yellow, cream to orange. Young animals and juveniles brighter than adults. Completely limbless.

DISTRIBUTION
Northeastern QLD, between Paluma and Thornton Peak.
HABITAT AND HABITS
Found in wet forests and rainforests. Cathemeral. Shelters beneath logs and rocks. Seen moving across trails and roads trails by day after rain and at night. Commonly mistaken for a small snake. Eats worms and other soft-bodied invertebrates. Thought to lay eggs.

S Eipper

Three-toed Snake-tooth Skink ■ *Coeranoscincus reticulatus* SVL 20cm

DESCRIPTION Body elongated. Tan to dark brown. Upper flanks greyish to purplish-black, becoming lighter towards tail. Usually a pale bar across snout. Lower flanks and underside yellow, cream to orange. Young animals and juveniles brighter than adults. Usually with cream or white head and forebody blotched or reticulated with black and grey. Short limbs with three digits on each.

DISTRIBUTION
Southeastern QLD, from Fraser Island, south to Wooli, NSW. **HABITAT AND HABITS** Found in wet forests and rainforests. Cathermeral. Shelters beneath logs and rocks or inside fallen epiphytes. Commonly mistaken for a small snake. Eats worms and other soft-bodied invertebrates. Lays 2–6 eggs per clutch.

S Eipper

Satinay Sand Skink ■ *Coggeria naufragus* SVL 13cm

DESCRIPTION Body elongated. Pale yellowish-brown, clear dorsolateral black stripe runs the length of body. Flanks and underside finely peppered with black. Four rudimentary limbs with three toes on each. **DISTRIBUTION** Restricted to Fraser Island, QLD.

HABITAT AND HABITS Lives in closed forests, rainforest and heaths on Fraser Island. Cathemeral. Fossorial. Rarely encountered, but found under rotting logs. Eats invertebrates, probably with a strong preference for earthworms due to the unusual dentition, where the teeth are pointed inwards towards the mouth, presumably to enable better grip on slippery prey. Mode of reproduction unknown.

Lemon-barred Skink ■ *Concinnia amplus* SVL 12cm

DESCRIPTION Robust build. Bronze-brown with dark grey flecks, which form an irregular series of thin cross-bands that darken on to tail. Usually black patch over shoulder. Flanks flecked with gold; lower flanks tinged with bright yellow. Underside bright yellow. Four limbs with five digits on each foot. **DISTRIBUTION** Northeastern QLD, between Prosperpine and Sarina. **HABITAT AND HABITS** Found in rainforests and neighbouring woodland. Diurnal. Shelters beneath logs and rocks. Unusual habit of sleeping exposed on vertical surfaces of boulders in rainforest streams. Most commonly seen while basking on rainforest trails. Eats invertebrates. Gives birth to 2–8 live young per litter.

Northern Bar-sided Skink ■ *Concinnia brachysoma* SVL 7.5cm

DESCRIPTION Body robust. Coppery-brown with black flecks and black blotches, which form irregular series of cross-bands that become more uniform on to tail. Underside grey.

Four limbs with five digits on each foot.
DISTRIBUTION Eastern QLD, from Gayndah to Artemis Station.
HABITAT AND HABITS Found in open woodland, rocky escarpments and urban environments. Diurnal. Often seen in and around houses. Shelters beneath logs and rocks. Eats invertebrates. Gives birth to 2–3 live young per litter.

S Eipper

Martin's Skink ■ *Concinnia martini* SVL 7cm

DESCRIPTION Body robust. Coppery-brown with black flecks and black blotches, which form irregular series of cross-bands that become more uniform on to tail. Underside grey.

Four limbs with five digits on each foot.
DISTRIBUTION Eastern QLD, from Mackay to Coffs Harbour, NSW.
HABITAT AND HABITS Found in open woodland, rocky escarpments and urban environments. Diurnal. Often seen in and around houses. Shelters beneath logs and rocks. Eats invertebrates; also consumes dried dog food and scavenges on food scraps. Gives birth to 2–4 live young per litter.

S Eipper

Bar-sided Skink ■ *Concinnia tenuis* SVL 9cm

DESCRIPTION Body robust. Silvery-grey to tan with black flecks and black blotches, which form an irregular series of cross-bands that become more uniform on to tail. These dark markings can be quite faded in some individuals. Underside grey. Four limbs with

five digits on each foot. **DISTRIBUTION** Eastern QLD, from Cairns to Narooma, NSW. **HABITAT AND HABITS** Found in forests, open woodland, rocky escarpments and urban environments. Diurnal. Often seen in and around houses. Shelters beneath logs and rocks. Eats invertebrates; also consumes dried dog food and scavenges on food scraps. Gives birth to 2–4 live young per litter.

Inland Snake-eyed Skink ■ *Cryptoblepharus australis* SVL 5cm

DESCRIPTION Body lightly built. Silver-grey to brown with black flecks and pair of ragged edged black paravertebral stripes. Pale dorsolateral stripes. Lower lateral dark-mottled grey to black. Underside cream. Four limbs with five digits on each foot. **DISTRIBUTION** Through inland Australia, from Moranbah, QLD, to Tennant Creek, NT, and south-west to Kalgoorlie, WA, across through inland SA and NSW. **HABITAT AND HABITS** Found in open woodland, mulga shrubland and urban environments. Diurnal. Often seen in and around houses. Shelters beneath logs and tree bark. Eats invertebrates. Lays 2–4 eggs per clutch.

Buchanan's Snake-eyed Skink ■ *Cryptoblepharus buchananii* SVL 5cm

DESCRIPTION Body lightly built. Silver-grey to brown with black flecks and pair of ragged edged black paravertebral stripes. Pale dorsolateral stripes. Lower lateral dark-mottled grey to black. Underside cream. Four limbs with five digits on each foot.

DISTRIBUTION Through WA, from Perth across to Kalgoorlie and north through inland WA, reaching coast between Onslow and Karatha. **HABITAT AND HABITS** Found in open woodland, shrubland and urban environments. Diurnal. Often seen in and around houses. Shelters beneath logs and tree bark. Eats invertebrates. Lays 2–4 eggs per clutch.

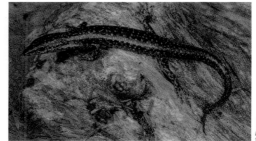

S Eipper

Christmas Island Blue-tailed Skink
■ *Cryptoblepharus egeriae* SVL 5cm

DESCRIPTION Body lightly built. Silver-grey with black flecks and pair of black paravertebral stripes. Pale dorsolateral stripes. Lower laterals dark-mottled grey to black. Tail metallic green to bright blue. Underside cream. Four limbs with five digits on each foot. **DISTRIBUTION** Formerly found on Christmas Island, where now extinct in the wild due to introduced predators, including the Asian Wolf Snake *Lycodon capucinus*. Recently introduced to the Cocos Keeling Islands. **HABITAT AND HABITS** Formerly seen across Christmas Island in forests, clearings and around settlements. Diurnal. Shelters beneath logs and tree bark. Eats invertebrates. Lays 1–3 eggs per clutch. Subject of successful captive-breeding programme.

Hamish Nodler

Coastal Snake-eyed Skink ■ *Cryptoblepharus littoralis* SVL 5cm

DESCRIPTION Body lightly built. Dark grey to brown with cream to black flecks. Lower lateral areas dark-mottled grey to black. Underside cream. Four limbs with five digits on each foot. **DISTRIBUTION** Nominate subspecies in coastal Queensland, between

Rockhampton and the Torres Strait islands. Additional subspecies *C. l. horneri* on the NT coastline, between Wessel Islands and Murgenella. **HABITAT AND HABITS** Found on beaches, mangroves, rocky shores and man-made structures such as piers. Diurnal. Shelters beneath logs and rocks, and inside cracks in both natural and man-made structures. Eats invertebrates. Lays 1–3 eggs per clutch.

Ragged Snake-eyed Skink ■ *Cryptoblepharus pannosus* SVL 5cm

DESCRIPTION Body lightly built. Dark grey to brown with black flecks and pair of ragged edged pale dorsolateral stripes. Lower lateral areas dark-mottled grey to black. Underside cream. Four limbs with five digits on each foot. **DISTRIBUTION** Eastern Australia, west of the GDR from Weipa, across to Cape Crawford, and south to Bordertown, SA, through

northern VIC and western NSW. **HABITAT AND HABITS** Found in open woodland, heaths, saltbush shrubland, mallee and urban environments. Diurnal. Shelters beneath logs and rocks, and inside cracks in both natural and man-made structures. Often seen on tree trunks, fences and buildings. Eats invertebrates. Lays 1–4 eggs per clutch.

Elegant Snake-eyed Skink ■ *Cryptoblepharus pulcher* SVL 5cm

DESCRIPTION Body lightly built. Silver-grey to brown with black flecks and pair of black paravertebral stripes. White dorsolateral stripes. Lower lateral areas dark-mottled grey to black. Tail metallic green to bright blue in juveniles in eastern NSW. Underside cream. Four limbs with five digits on each foot. **DISTRIBUTION** Nominate subspecies up the eastern seaboard from Milton, NSW, to Ingham, QLD; subspecies C. *p. clarus* across southern Australia between Yorke Peninsula, SA, and Cape Arid, WA. **HABITAT AND HABITS** Lives in open woodland, heaths, saltbush shrubland, mallee and urban environments. Diurnal. Shelters beneath logs and rocks, and inside cracks, in both natural and man-made structures. Often seen on tree trunks, fences and buildings. Eats invertebrates. Lays 1–4 eggs per clutch.

S Eipper

Ariadna's Skink ■ *Ctenotus ariadnae* SVL 6.5cm

DESCRIPTION Body medium in build. Dark brown to black with series of fine, whole and broken, yellow stripes that become orange over hips and on to tail. Underside creamy-white. Four limbs with five digits on front feet and five on rear feet. **DISTRIBUTION** Three populations, one around the Laverton region of WA, between Telfer and north-west SA border; others on the three-way border region of QLD, south-east NT and north-east SA. **HABITAT AND HABITS** Found on dune systems and spinifex grassland. Diurnal. Shelters in burrows dug beneath clumps of vegetation, logs and rocks. Eats invertebrates. Lays 3–6 eggs per clutch.

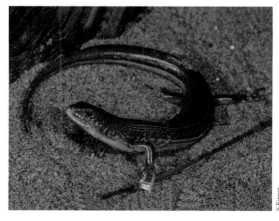

S Eipper

Brown-backed Skink ■ *Ctenotus coggeri* SVL12 cm

DESCRIPTION Body medium in build. Head yellowish. Body plain, pale to dark brown, with fine, gold-edged dorsolateral stripe that borders broad brown to black lateral stripe.

Lower white lateral stripe along body. Underside bright yellow. Four limbs with five digits on front feet and five on rear feet. **DISTRIBUTION** Western Arnhem Land Escarpment, NT. **HABITAT AND HABITS** Found in rocky gorges and tropical open woodland on stony soils. Diurnal. Shelters beneath logs and rocks. Eats invertebrates. Lays 2–8 eggs per clutch.

West Coast Skink ■ *Ctenotus fallens* SVL 12cm

DESCRIPTION Body medium in build. Pale to dark brown, usually with pale-edged black vertebral stripe. White-edged dorsolateral stripe with series of white spots over broad

brown lateral stripe. Underside creamy-white. Four limbs with five digits on front feet and five on rear feet. **DISTRIBUTION** Coastal southern WA, between Carnarvon and Bunbury. **HABITAT AND HABITS** Found in woodland, heaths, grassland and urban centres. Diurnal. Shelters in burrows dug beneath clumps of vegetation, logs and rocks. Eats invertebrates. Lays 3–10 eggs per clutch.

Red-legged Skink ▪ *Ctenotus labillardieri* SVL 7.5cm

DESCRIPTION Body medium in build. Pale to dark brown, usually with pale-edged black vertebral stripe. White-edged dark dorsolateral stripe above broad black lateral stripe that contains white spots. Lower lateral white stripe. Underside creamy-white. Four reddish-

brown limbs with five digits on front feet and five on rear feet. **DISTRIBUTION** Southern WA, between Perth and Cape Arid. **HABITAT AND HABITS** Found in woodland, heaths, grassland and urban centres. Diurnal. Shelters in burrows dug beneath clumps of vegetation, logs and rocks. Eats invertebrates. Lays 2–5 eggs per clutch.

M Summerville

Gravelly-soil Ctenotus ▪ *Ctenotus lateralis* SVL 8.5cm

DESCRIPTION Body medium in build. Reddish-brown with series of black, grey and white stripes. Underside and lower lateral zone creamy-white. Four limbs with five digits on front feet and five on rear feet. **DISTRIBUTION** Western QLD, centred on the Selwyn Range

between Doomagee to Winton. Isolated population between Cobbold Gorge and Chillagoe. **HABITAT AND HABITS** Found in rocky gorges, spinifex grassland and open woodland, usually on stony soils. Diurnal. Shelters in burrows dug beneath clumps of vegetation, logs and rocks. Eats invertebrates. Lays seven eggs per clutch.

S Eipper

Common Desert Ctenotus ■ *Ctenotus leonhardii* SVL 8cm

DESCRIPTION Body medium in build. Reddish-brown with series of black and dark brown stripes dorsally. White dorsolateral stripe. Sides reddish-brown with complex pattern of fine white spots and series of larger white spots that coalesce into streaks. Underside creamy-white. Four limbs with five digits on front feet and five on rear feet.

DISTRIBUTION Arid central Australia, from Charleville, QLD, across to the Gascoyne region of WA. **HABITAT AND HABITS** Found in dune system spinifex grassland and open woodland, usually on sandy soils. Diurnal. Shelters in burrows dug beneath clumps of vegetation, logs and rocks. Eats invertebrates. Lays 1–7 eggs per clutch.

Eastern Spotted Ctenotus ■ *Ctenotus orientalis* SVL 8cm

DESCRIPTION Body medium in build. Brown with black vertebral stripe edged in gold. Row of white spots between dorsolateral stripe and paravertebral stripes. Sides black with white flecks and spots. Tail brown with yellow-orange flush. Underside creamy-white. Four limbs with five digits on front feet and five on rear feet. **DISTRIBUTION** Southern

Australia, between Balladonia, WA, across SA, and VIC to Lake Wyangala, NSW. **HABITAT AND HABITS** Found in dune systems, mallee, spinifex grassland and open woodland, usually on sandy soils. Diurnal. Shelters in burrows dug beneath clumps of vegetation, logs and rocks. Eats invertebrates. Lays 3–7 eggs per clutch.

Leopard Ctenotus ■ *Ctenotus pantherinus* SVL 9cm

DESCRIPTION Body robust. Reddish to dark brown with series of black flecks and white spots. Underside creamy-white. Four limbs with five digits on front feet and five on rear feet. A species complex. **DISTRIBUTION** Much of Australia, from St George, QLD, west through the NT, SA, and across WA north of Brookton. **HABITAT AND HABITS** Found in dune systems, spinifex grassland and open woodland, usually on sandy soils. Other populations found on rocky soils. Diurnal but also active on warm nights. Shelters in burrows dug beneath clumps of vegetation, logs and rocks. A number of subspecies are described in its range. Eats invertebrates. Lays 3–11 eggs per clutch.

S Eipper

Pretty Ctenotus ■ *Ctenotus pulchellus* SVL 8cm

DESCRIPTION Body medium in build. Pale brown with series of fine black stripes across back. Thicker black dorsolateral stripe above bright orange-red flanks that are flecked with white. Underside creamy-white. Four limbs with five digits on front feet and five on rear feet. **DISTRIBUTION** Western QLD, from Winton across to Tennant Creek, NT. **HABITAT AND HABITS** Found in rocky gorges, spinifex grassland, black-soil plains and open woodland. Diurnal. Shelters under clumps of vegetation, logs and rocks. Eats invertebrates. Lays 4–7 eggs per clutch.

R Francis

Royal Ctenotus ■ *Ctenotus regius* SVL 8cm

DESCRIPTION Body medium in build. Reddish-brown with pale-edged black vertebral stripe. Dark-edged white dorsolateral stripe above dark sides covered in white spots and blotches. Lower lateral white stripe. Tail brown with yellow-orange flush; each side of original tail has continuous black stripe. Underside creamy-white. Four limbs with five digits on front feet and five on rear feet. **DISTRIBUTION** Southern central Australia, between Madura, WA, across through SA, north-west VIC, western NSW, to Dalby, QLD. **HABITAT AND HABITS** Found in dune systems, mallee, brigalow and open woodland, usually on sandy soils. Diurnal. Shelters in burrows dug beneath clumps of vegetation, logs and rocks. Eats invertebrates. Lays 1–4 eggs per clutch.

R Francis

Eastern Striped Ctenotus ■ *Ctenotus robustus* SVL 12cm

DESCRIPTION Body medium to robust. Pale to dark brown, usually with pale-edged black vertebral stripe. White-edged dorsolateral stripe with series of white spots over broad brown lateral stripe. Plain form in coastal regions lacks typical dark stripes. Underside creamy-white. Four limbs with five digits on front feet and five on rear feet.

S Eipper

A species complex. **DISTRIBUTION** Across Australia, except southern WA, western SA and TAS. **HABITAT AND HABITS** A true generalist, found in deserts, woodland, heaths, grassland and urban centres. Diurnal. Shelters in burrows dug beneath clumps of vegetation, logs and rocks. Eats invertebrates. Lays 2–11 eggs per clutch.

Copper-tailed Skink ■ *Ctenotus taeniolatus* SVL 9cm

DESCRIPTION Body medium in build. Black with series of white to pale yellow stripes. Tail yellow to coppery-orange. Underside creamy-white. Four limbs with five digits on front feet and five on rear feet. Probably a species complex. **DISTRIBUTION** Eastern Australia, between East Gippsland up east coast, through eastern NSW and QLD, to Edmonton. **HABITAT AND HABITS** Found in woodland, heaths, grassland and open woodland, usually with rocky outcrops. Diurnal. Shelters in burrows dug beneath clumps of vegetation, logs and rocks. Eats invertebrates. Lays 2–7 eggs per clutch.

S Eipper

Tasmanian She-oak Skink ■ *Cyclodomorphus casuarinae* SVL 18cm

DESCRIPTION Body elongated. Tan, reddish-brown to silvery-grey, with or without black flecks and streaks running along body. Often small, bright yellow, orange and red flecks along sides. Usually dark marking over eyes, sometimes extending towards ear opening. All-black patternless individuals occasionally occur. Juveniles heavily banded. Underside usually strongly marked with yellow, orange, red and black patterning. Four short, muscular limbs with five digits on each foot. **DISTRIBUTION** Across mainland TAS. **HABITAT AND HABITS** Found in woodland, heaths and tussock grassland. Diurnal. Shelters beneath leaf litter and in top layers of soil. Sometimes uncovered beneath logs and rocks. When threatened, opens mouth and flicks out its blue tongue rapidly. Eats invertebrates, snails and vegetation. Gives birth to 2–11 live young per litter.

S Eipper

Western Slender Blue-tongue ■ *Cyclodomorphus celatus* SVL 12cm

DESCRIPTION Body elongated. White to silvery-grey with thin black flecks and streaks running along body. Often small, bright yellow, orange and red flecks along sides. Underside yellow with dark and light flecking. Four short, muscular limbs with five digits

on front feet and five on rear feet. **DISTRIBUTION** Along the WA coastline, between Perth and Carnavon. **HABITAT AND HABITS** Found in banksia woodland, heaths, grassland and urban centres. Diurnal. Shelters beneath leaf litter and in top layers of soil. Sometimes uncovered beneath logs, rocks or rubbish sitting on sand. Eats invertebrates, snails and vegetation. Gives birth to 2–6 live young per litter.

Pink-tongued Skink ■ *Cyclodomorphus gerrardii* SVL 20cm

DESCRIPTION Body elongated. Tan, reddish-brown to silvery-grey, with or without dark flecks and streaks running along body. Usually broad bands of chocolate-brown to dark grey, generally black in juveniles on body and tail. Generally dark marking over eyes, sometimes extending towards ear opening. Completely patternless individuals are known. Underside usually like back. Four short, muscular limbs with five digits on each foot. **DISTRIBUTION** Across eastern Australia, along the GDR from Menai, NSW, to

Port Douglas, QLD. **HABITAT AND HABITS** Found in rainforest, woodland, heaths and urban centres. Shelters beneath leaf litter and in top layers of soil. Mainly nocturnal. Often seen after rain. When threatened, opens mouth and flicks out the usually pink tongue rapidly. Eats invertebrates, snails and vegetation. Gives birth to 2–29 live young per litter.

Spinifex Slender Blue-tongue ■ *Cyclodomorphus melanops* SVL 13cm

DESCRIPTION Body elongated. Grey to dark olive-brown with thin black flecks on each scale. Young and subadults usually have yellow to white spots across body, but more numerous on flanks. Underside cream to yellow with dark and light flecking. Four short, muscular limbs with five digits on each foot. Probably a species complex, with four subspecies. **DISTRIBUTION** Disjunct populations across much of arid Australia. *C. m. melanops* through WA and central NT; *C. m. elongatus* eastern and central SA, south-west NSW and western QLD; *C. m. siticulosus* across the Eyre Peninsula and Nullarbor Plain of SA and WA.

HABITAT AND HABITS Diurnal, but nocturnal in warmer weather. Found in open woodland, saltbush shrubland, grassland and sandy deserts with spinifex. Shelters beneath leaf litter and in top layers of soil. Sometimes uncovered beneath logs, rocks or rubbish sitting on sand. Eats invertebrates, snails and vegetation. Gives birth to 2–7 live young per litter.

H Cogger

Mainland She-oak Skink ■ *Cyclodomorphus michaeli* SVL 12cm

DESCRIPTION Body elongated. Tan, reddish-brown to silvery-grey, with or without black flecks and streaks along body. Small, bright yellow, orange and red flecks along sides. Usually dark marking over eye, sometimes extending towards ear opening. Underside strongly marked with yellow, orange, red and black patterning. Four short, muscular limbs with five digits on each foot. **DISTRIBUTION** Southeastern Australia, in disjunct populations from east Gippsland, VIC, to Bega, NSW, around Wollongong to Newcastle, NSW, and in the New England Tableland around Armidale.

HABITAT AND HABITS Found in woodland, heaths, tussock grassland and urban fringes. Diurnal. Shelters beneath leaf litter and in top layers of soil. Sometimes uncovered beneath logs, rocks or rubbish sitting on soil. When threatened, flicks out the blue tongue rapidly. Eats invertebrates, snails and vegetation. Gives birth to 2–11 live young per litter.

J Meney

Cunningham's Skink ▪ *Egernia cunninghami* SVL 18cm

DESCRIPTION Body robust. Highly variable, from pale brown to black, with or without irregular pale or darker markings often flecked with white, yellow or orange. Underside in most populations pinkish-orange with dark and light flecking. Flattened body with spine-like, keeled scales that vary among the populations that are suited to the shelter type they utilize. Four muscular limbs with five digits on each foot. A species complex requiring further work to resolve. **DISTRIBUTION** From Carnarvon Gorge NP, QLD, south along the GDR, through NSW, on to the basalt plains of VIC. Isolated population on the Fleurieu

Peninsula and Adelaide Hills, SA. **HABITAT AND HABITS** Found in woodland and grassland with rocky outcrops. Diurnal. Shelters inside rock crevices, under rocks and rarely in trees. Mainly eats vegetation but will also eat invertebrates. Crevices used by family group are given away by presence of shared latrine site. Gives birth to 2–19 live young per litter.

Southern Pygmy Spiny-tailed Skink ▪ *Egernia depressa* SVL 10cm

DESCRIPTION Body robust. Head and body anteriorly reddish-orange to brown, with irregular pale grey markings; posterior grey with irregular reddish-orange blotches. Short, spiky tail. Underside in most populations whitish-grey with darker flecking. Flattened body with spiny scales that are used to help wedge it into crevices inside dead mulga trees. Four muscular limbs with five digits on each foot. **DISTRIBUTION** From North

West Cape, across to Newman, and south to Kalgoorlie and Dowerin, WA. **HABITAT AND HABITS** Occurs in mallee, mulga and other open woodland. Diurnal. Usually found in hollow trees, indicated by shared latrine site at a tree's base. Mainly eats vegetation but will also eat invertebrates. Gives birth to 1–5 live young per litter.

Eastern Pilbara Spiny-tailed Skink ■ *Egernia epsilosus* SVL 11cm

DESCRIPTION Body robust. Reddish-orange with pale orange bands. On hips and tail small black blotches form irregular bands. Short, spiky tail. Underside in most populations whitish with orange flecking. Flattened body with spiny scales that are used to help wedge it into rock crevices. Four muscular limbs with five digits on each foot. **DISTRIBUTION** Eastern Pilbara, between Shay Gap and Mt Francisco, WA. **HABITAT AND HABITS** Found in rocky gorges with woodland. Diurnal. Usually occurs in rock piles; shelter site used indicated by common latrine site at base of rock face. Mainly eats vegetation but will also eat invertebrates. Gives birth to 1–6 live young per litter.

S Eipper

Hosmer's Skink ■ *Egernia hosmeri* SVL 18cm

DESCRIPTION Body robust. Chocolate to yellowish-brown with darker flecking. Large white to yellow spots cover body, along with darker flecks. Underside pinkish-orange with dark and light flecking. Spine-like keeled scales. Flattened body. Four muscular limbs with five digits on each foot; soles of feet black. **DISTRIBUTION** Mt Surprise, across northern QLD, to Macarthur River, NT. **HABITAT AND HABITS** Found in woodland, spinifex desert and grassland with rock outcrops. Diurnal. Shelters inside rock crevices, under rocks and rarely in trees. Mainly eats vegetation but will also eat invertebrates. Crevice location used by family group indicated by presence of shared latrine site. Gives birth to 2–10 live young per litter.

S Eipper

King Skink ■ *Egernia kingii* SVL 25cm

DESCRIPTION Body robust. White to black with or without thin, dark and light flecks and streaks along body. Underside yellowish to orange with dark and light flecking. Strongly keeled scales. Four muscular limbs with five digits on front feet and five on rear

feet. **DISTRIBUTION** WA coastline, between Shark Bay and Duke of Orleans Bay. **HABITAT AND HABITS** Found in woodland, heaths, grassland and urban environments. Diurnal. Shelters in burrows dug under rocks and logs. Mainly eats invertebrates and vegetation but will also scavenge on food scraps. Gives birth to 2–14 live young per litter.

Eastern Crevice Skink ■ *Egernia mcpheei* SVL 14cm

DESCRIPTION Body robust. Charcoal-grey to brown with darker flecking. Broad dark line extends from head-sides, along flanks and on to hips. Whole upper surface has fine white speckles across it. Underside pinkish-orange with dark and light flecking. Strongly keeled scales. Flattened body. Four muscular limbs with five digits on each foot; soles of feet

black. **DISTRIBUTION** NSW, from Barrington Tops along the GDR to the McPherson Range, QLD. **HABITAT AND HABITS** Found in woodland, heaths, grassland and urban centres. Diurnal. Shelters inside rock crevices, under rocks and rarely in trees. Mainly eats invertebrates but will occasionally eat fruits and berries. Gives birth to 3–6 live young per litter.

South-western Crevice Skink ■ *Egernia napoleonis* SVL 13cm

DESCRIPTION Body robust. Silver-grey to brown with dark, irregular vertebral stripe. Dark line extends from head-sides, along flanks and on to hips. Fine white speckles across whole upper surface. Underside pinkish-orange with dark and light flecking. Strongly keeled scales. Flattened body. Four muscular limbs with five digits on each foot.

DISTRIBUTION Along WA coastline, between Lancelin and Cape Arid.
HABITAT AND HABITS Found in woodland, heaths, grassland and urban environments. Diurnal. Shelters in dead grass, tree trunks and rock crevices, and under rocks. Mainly eats invertebrates and berries. Gives birth to 1–7 live young per litter.

Yakka Skink ■ *Egernia rugosa* SVL 24cm

DESCRIPTION Body robust. Yellowish-brown to dark brown above. Centre of back greyish-black. Yellow dorsolateral line borders brown upper lateral side. Lower lateral yellowish. Underside yellowish to orange with dark and light flecking. Strongly keeled scales. Four muscular limbs with five digits on front feet and five on rear feet.
DISTRIBUTION Much of eastern QLD, from Bollon through the brigalow to Charters

Towers. Isolated populations near Etheridge and on the southern Cape York Peninsula near Aurukun. **HABITAT AND HABITS** Found in open woodland, brigalow and grassland. Diurnal. Shelters in burrows dug under rocks and logs. Mainly eats vegetation and invertebrates. Gives birth to 2–8 live young per litter.

Black Rock Skink ■ *Egernia saxatilis* SVL 13cm

DESCRIPTION Body robust. Charcoal-grey to brown with darker flecking. Broad dark line extends from head-sides, along flanks and on to hips. Fine white speckles across whole upper surface. Underside pinkish-orange with dark and light flecking. Strongly keeled scales. Flattened body. Four muscular limbs with five digits on feet; soles of feet black. **DISTRIBUTION** Two subspecies. *E. s. saxatilis* in central NSW around the

Warrumbungles; *E. s. intermedia* from the Grampians, through VIC and south-east NSW, to the Blue Mountains. **HABITAT AND HABITS** Found in woodland, heaths, grassland and urban centres. Diurnal. Shelters inside rock crevices, under rocks and rarely in trees. Mainly eats invertebrates but will occasionally eat fruits and berries. Gives birth to 2–9 live young per litter.

Tree Skink ■ *Egernia striolata* SVL 12cm

DESCRIPTION Body robust. Pale to charcoal-grey; can also be brown with darker flecking. Broad dark line extends from head-sides, along flanks and on to hips. Fine white speckles across whole upper surface. Underside pale yellow or orange with dark and light flecking. Strongly keeled scales. Flattened body. Four muscular limbs with five digits on each foot; soles of feet brown. Probably a species complex. **DISTRIBUTION** From outskirts of

Adelaide, SA, throught north-west VIC, and western NSW and QLD, to Charters Towers. **HABITAT AND HABITS** Found in open woodland, brigalow, mallee, grassland and urban centres. Diurnal. Shelters inside rock crevices, under rocks, and commonly in trees both fallen and standing. Mainly eats invertebrates but will occasionally eat fruits and berries. Gives birth to 2–7 live young per litter.

Gidgee Skink ■ *Egernia stokesii* SVL 18cm

DESCRIPTION Body robust. Highly variable, from pale brown to black, with or without irregular pale or darker markings; often flecked with white, yellow or orange. Underside in most populations pinkish-orange with dark and light flecking. Short, spiny tail. Four muscular limbs with five digits on each foot. Probably a species complex. **DISTRIBUTION** Multiple isolated populations. *E. s. stokesii* in the Houtman Abrolhos island group; *E. s. badia* in south-west inland WA, from the Gascoyne region to the Eastern Goldfields;

E. s. zellingi in eastern SA and western NSW, to Longreach, QLD. **HABITAT AND HABITS** Found in open woodland, Mitchell grassland, rocky gorges, hummock grassland and brigalow. Diurnal. Some populations live in rock crevices, under rocks, and others in hollow trees. Mainly eats vegetation but will also eat invertebrates. Gives birth to 1–8 live young per litter.

T Eipper

Eastern Narrow-banded Sandswimmer
■ *Eremiascincus fasciolatus* SVL 12cm

DESCRIPTION Body medium in build. Yellow with 10–16 purple to black bands. Tail has 25–33 bands. Underside creamy-white. Four limbs with five digits on front feet and five on rear feet. **DISTRIBUTION** Eastern QLD, between Purga and Mitchell, north to Moranbah. **HABITAT AND HABITS** Found in open woodland, grassland and brigalow, usually on sandy soils. Diurnal to crepuscular. Shelters in burrows dug beneath clumps of vegetation, logs and rocks. Eats invertebrates. Lays 3–9 eggs per clutch.

S Eipper

Broad-banded Sandswimmer ■ *Eremiascincus richardsonii* SVL 10cm

DESCRIPTION Body medium in build. Yellow with 8–14 purple to black bands. Tail has 19–32 bands. Underside creamy-white. Four limbs with five digits on front feet and five on

rear feet. Probably a species complex. **DISTRIBUTION** Central Australia, from western QLD and NSW, across through SA and NT, to WA coastline between Perth and Broome. **HABITAT AND HABITS** Found in dune systems, mallee, spinifex grassland and open woodland, usually on sandy soils. Diurnal to crepuscular. Shelters in burrows dug beneath clumps of vegetation, logs and rocks. Eats invertebrates. Lays 3–11 eggs per clutch.

Elf Skink ■ *Eroticoscincus graciloides* SVL 3cm

DESCRIPTION Body elongated. Pale brown to dark brown with pale flecks. Strongly pointed snout. Lips barred with white. Sometimes a rusty dorsolateral stripe over hips extending on to tail. Underside cream to white. Very shiny scales refract in sunlight, giving rainbow-like appearance. Four thin limbs with four digits on front feet and five on rear feet. **DISTRIBUTION** Between Pine Mountain and Fraser Island, QLD. **HABITAT AND HABITS** Found in closed forest, vine thickets, rainforest, and some well-watered urban gardens in Brisbane. Diurnal. Fossorial. Shelters in leaf litter, and under fallen timber and rocks. Mainly eats invertebrates. Presumed to be egg laying.

Brown-sheen Skink ■ *Eugongylus rufescens* SVL 17cm

DESCRIPTION Body elongated and robust. Reddish-brown to silvery-grey. Faint barring on lips in adults; more prominent in young animals. Juveniles reddish-brown with fine white to yellow bands that cover body and tail. Scales have opalescent appearance, giving purplish appearance in bright light. Four short, muscular limbs with five digits on each foot. **DISTRIBUTION** QLD, on tip of Cape York near Lockerbie Scrub, and on many Torres Strait Islands. Also PNG and Indonesia. **HABITAT AND HABITS** Found in rainforest, woodland and heaths. Diurnal. Fossorial. Shelters beneath leaf litter and inside rotting logs. When threatened, opens mouth and flicks out pink tongue rapidly. Eats spiders, insects and smaller lizards. Lays 1–5 eggs per clutch.

Adult

Juvenile

Yellow-bellied Water Skink ■ *Eulamprus heatwolei* SVL 10cm

DESCRIPTION Body robust. Bronze to coppery-brown. Pattern of irregular black flecks and spots scattered across body and tail. Upper flanks black with fine white spots; lower flanks pale grey to whitish, flecked with black towards underside. Underside pale yellow with black flecks. Four muscular limbs, each with five digits on foot. Probably a species complex. **DISTRIBUTION** East-coast highlands from Ebor, NSW, along the GDR, to south-east NSW and Gippsland, to Seymour, VIC. Isolated population on the Fleurieu Peninsula, SA. **HABITAT AND HABITS** Found in swamps, creeklines and rivers, and around lakes; also commonly in urban environments. Diurnal. Shelters under rocks, logs and man-made debris. Usually seen basking near cover. Mainly eats invertebrates and small vertebrates. Gives birth to 1–6 live young per litter.

Blue Mountains Water Skink ■ *Eulamprus leuraensis* SVL 13cm

DESCRIPTION Body robust. Dark brown to black, mottled with bright yellow markings and fine white to yellow dots. Copper-coloured head heavily marked with black. Series of fine pale paravertebral and dorsolateral stripes extends on to tail. Upper flanks black; lower flanks heavily spotted with yellow and bronze. Underside yellow with black spots.

Markings more prominent in juveniles than adults. Four muscular limbs with five digits on each foot. **DISTRIBUTION** Between Wentworth Falls and Lithgow, NSW. **HABITAT AND HABITS** Found in swamps and creeklines, and around lakes. Diurnal. Shelters in holes, under rocks and logs, and man-made debris. Usually seen basking near cover or on top of grass tussocks. Mainly eats invertebrates. Gives birth to 2–5 live young per litter.

Eastern Water Skink ■ *Eulamprus quoyii* SVL 12cm

DESCRIPTION Body robust. Bronze to coppery-brown with prominent golden-yellow dorsolateral stripe. Pattern of irregular black flecks scattered across body and on to tail. Upper flanks black with fine white spots; lower flanks pale grey to whitish, flecked with black towards underside. Underside white. Four muscular limbs with five digits on each foot. **DISTRIBUTION** Between Mossman, QLD, along east coast to Batemans Bay. Also along the Murray/Darling drainage to north-west VIC, south-west NSW, to Adelaide, SA. **HABITAT AND HABITS** Found in swamps, creeklines and rivers, and around lakes and urban environments. Diurnal. Shelters under rocks, logs and man-made debris. Usually seen basking near cover. Mainly eats invertebrates and small vertebrates. Gives birth to 2–8 live young per litter.

Southern Water Skink ■ *Eulamprus tympanum* SVL 10cm

DESCRIPTION Body robust. Bronze to coppery-brown. Pattern of irregular black flecks and spots scattered across body and tail. Upper flanks black with fine white spots; lower flanks pale grey to whitish, flecked with black towards underside. Underside pale yellow with black flecks but can be heavily marked with black. Four muscular limbs with five digits on each foot. Subspecies *E. t. marnae* found in south-west VIC is probably nothing more than a colour variant.

DISTRIBUTION Eastern coast from the Blue Mountains, through south-east NSW and southern VIC, to Mt Gambier, SA. **HABITAT AND HABITS** Found in swamps, creeklines and rivers, and around lakes; also common in urban environments. Diurnal. Shelters under rocks, logs and man-made debris. Usually seen basking near cover. Mainly eats invertebrates and small vertebrates. Gives birth to 1–7 live young per litter.

S. Eipper

Atherton Tableland Mulch-skink
■ *Glaphyromorphus mjobergi* SVL 10cm

DESCRIPTION Body elongated. Bronze-brown to dark purplish-grey with fine irregular black flecks. Colouration becomes darker on tail. Yellow to cream dorsolateral stripe. Flanks blackish-grey with fine white flecks. Underside white. Four muscular limbs with five digits on each foot. **DISTRIBUTION** Above 650m in north-east QLD, between Tully Falls and Mt Carbine. **HABITAT AND HABITS** Found in closed forest and rainforest. Diurnal. Shelters in rotten logs and under fallen timber and rocks. Mainly eats invertebrates. Lays three eggs per clutch.

S. Eipper

Fine-spotted Mulch-skink ■ *Glaphyromorphus punctulatus* SVL 7cm

DESCRIPTION Body elongated. Pale brown to dark brown with fine irregular dark and

pale flecking. Colouration lighter on lower flanks. Underside white. Four thin limbs with five digits on each foot. **DISTRIBUTION** Eastern QLD, between Cairns, along the GDR, to Gayndah. **HABITAT AND HABITS** Found in closed forest and open woodland. Diurnal. Fossorial. Shelters under fallen timber and rocks. Mainly eats invertebrates. Lays three eggs per clutch.

Prickly Forest Skink ■ *Gnypetoscincus queenslandiae* SVL 9cm

DESCRIPTION Body robust with rough scales. Dark brown with series of lighter, thin irregular bands. Underside bright yellow with brown spotting. Four robust limbs with five digits on each foot. **DISTRIBUTION** Between Kirrama along east coast to Rossville, QLD. **HABITAT AND HABITS** Found in rainforest. Diurnal. Shelters in leaf litter, and under fallen timber and rocks. Usually seen with head protruding from rotten log or while moving on rainforest floor, particularly after rain, out of direct sunlight. Mainly eats invertebrates. Gives birth to 2–6 live young per litter.

Rainforest Cool-skink ■ *Harrisoniascincus zia* SVL 5cm

DESCRIPTION Body robust. Pale brown to dark brown with pale dorsolateral line. Flanks dark, becoming lighter towards underside. Whole upper body flecked with pale brown. Underside bright yellow to white. Four thin limbs with five digits on each foot. **DISTRIBUTION** Between Canungra, QLD, along east coast to Dorrigo, NSW. **HABITAT AND HABITS** Found in beech forest and rainforest. Diurnal. Shelters in leaf litter, and under fallen timber and rocks. Usually seen basking on edges of walking trails in forests. Mainly eats invertebrates. Egg laying.

R Francis

Three-toed Earless Skink ■ *Hemiergis decresiensis* SVL 5cm

DESCRIPTION Body elongated. Pale to dark brown or grey, usually with 2–4 thin black stripes that run along body, starting on neck and running on to tail. Underside bright yellow. Four short, thin limbs with three digits on each foot. **DISTRIBUTION** Central Victoria, from Inglewood, through western VIC and eastern SA, to Kangaroo Island. **HABITAT AND HABITS** Lives in open woodland and rocky slopes. Cathemeral. Fossorial. Shelters beneath logs, rocks and man-made debris; usually seen disappearing into the soil when uncovered. Eats invertebrates. Gives birth to 2–5 live young per litter.

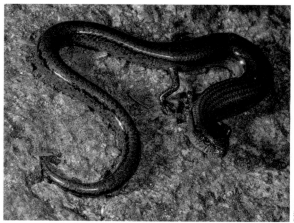

S Eipper

Two-toed Earless Skink ■ *Hemiergis quadrilineata* SVL 7cm

DESCRIPTION Body elongated. Pale to dark brown or grey, generally with two thin black stripes that run along body, starting on neck and running on to tail. Usually dark stripe along each flank. Underside bright yellow. Four short, thin limbs with two digits on

each foot. **DISTRIBUTION** Coastal WA, from Yallingup, north to Cervantes; very common in Perth. **HABITAT AND HABITS** Lives in open woodland, swamps, heaths and urban areas. Nocturnal. Fossorial. Shelters beneath logs, rocks and man-made debris; usually seen disappearing into the soil when uncovered. Commonly mistaken for a small snake. Eats invertebrates. Gives birth to 2–6 live young per litter.

Murray's Skink ■ *Karma murrayi* SVL 11cm

DESCRIPTION Body robust. Dark brown to dark grey with fine irregular black bars, becoming darker on tail. Flanks blackish-grey with fine white to pale blue flecks and yellow spots. Underside pale yellow. Four muscular limbs with five digits on each foot. **DISTRIBUTION** Between the Conondale Range, QLD, along east coast to Gloucester, NSW. **HABITAT AND HABITS** Found in closed forest and rainforest. Diurnal. Shelters in rotten logs and under fallen timber and rocks. Usually seen basking on edges of rotten logs along walking trails in forests. Mainly eats invertebrates. Gives birth to 2–5 live young per litter.

Rainforest Sunskink ■ *Lampropholis coggeri* SVL 6cm

DESCRIPTION Body medium in build. Dark brown to brown, generally with grey flecking that becomes more prominent towards tail. Bronze dorsolateral stripe, usually with darker flanks. Four limbs with five digits on each foot. **DISTRIBUTION** Northeastern QLD, between Lake Barrine and Cooktown. **HABITAT AND HABITS** Found in rainforest, woodland and urban environments. Diurnal. Shelters beneath logs and rocks. Most commonly seen basking on fallen branches and among leaf litter. Eats invertebrates. Lays 1–4 eggs per clutch.

M Sanders/ Ecosmart Ecology

Garden Skink ■ *Lampropholis delicata* SVL 6cm

DESCRIPTION Body medium in build. Dark brown to brown, generally with grey flecking that becomes more prominent towards tail. Bronze dorsolateral stripe, usually with darker flanks. Sometimes a white lower lateral stripe. Four limbs with five digits on each foot. **DISTRIBUTION** Eastern Australia, between Port Douglas, QLD, and Geelong, VIC; isolated populations in south-west VIC, Adelaide Hills, southern Eyre Peninsula, SA, and TAS. Introduced populations in Hawaii and New Zealand. **HABITAT AND HABITS** Found in rainforest, heaths, woodland and urban environments. Diurnal. Shelters beneath logs and rocks. Most commonly seen basking in gardens. Eats invertebrates. Lays 1–4 eggs per clutch, often communally. Communal egg sites can have thousands of eggs.

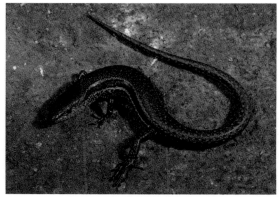

T Eipper

Grass Skink ▪ *Lampropholis guichenoti* SVL 7cm

DESCRIPTION Body medium in build. Head usually copper coloured. Body silver to greyish-brown with black vertebral stripe, generally with grey and white flecking. Bronze dorsolateral stripe, usually with darker flanks and lower lateral stripe. Four limbs with five digits on each foot. **DISTRIBUTION** Eastern Australia, between Inskip Point, QLD, and

Mt Gambier, SA. Isolated population in SA between Kangaroo Island and Port Augusta. **HABITAT AND HABITS** Found in heaths, coastal dunes, woodland and urban environments. Diurnal. Shelters beneath logs and rocks. Most commonly seen basking in gardens. Eats invertebrates. Lays 1–5 eggs per clutch, often communally. Communal egg sites can have thousands of eggs.

S Eipper

Western Two-toed Slider ▪ *Lerista bipes* SVL 7cm

DESCRIPTION Body elongated. Pale yellow to reddish-brown, becoming greyish on to tail. Two thin black stripes formed by flecks run along body, starting on neck and running on to tail. Thick dark stripe along each flank. Underside white. Two short, thin rear limbs with two digits on each foot. **DISTRIBUTION** Much of central Australia, between Quilpie, south-west QLD, through the NT north to Elliott, across into the Kimberley of

WA, south to Carnarvon, south-east to Kitchener and into northern SA. **HABITAT AND HABITS** Found in dune systems, spinifex grassland and open woodland, usually on sandy soils. Diurnal. Fossorial. Shelters in litter beneath clumps of vegetation and trees, logs and rocks. Eats invertebrates. Lays 1–4 eggs per clutch.

A Zimny

Bouganville's Slider ■ *Lerista bougainvillii* SVL 7cm

DESCRIPTION Body elongated. Pale to dark brown or grey, usually with fine black flecking across body. Original tail often has yellow to red flush with fine brown reticulations. Thick dark stripe along each flank. Underside bright yellow. Four short, thin limbs with five digits on each foot. **DISTRIBUTION** From the Eyre Peninsula, through the Flinders Ranges, SA, across most of VIC, and on to the GDR and western slopes of NSW.

Also Flinders Island and mainland TAS. **HABITAT AND HABITS** Lives in open woodland and on rocky slopes. Nocturnal. Fossorial. Shelters beneath logs, rocks and man-made debris; usually seen disappearing into the soil when uncovered. Commonly mistaken for a small snake. Eats invertebrates. Lays 1–4 eggs per clutch.

R Francis

Myall Slider ■ *Lerista edwardsae* SVL 8cm

DESCRIPTION Body elongated. Pale yellow to reddish-brown, becoming greyish on to tail. Two thin black stripes along body, starting on neck and extending on to tail. Regenerated tails lack stripes. Small black flecks across body. Thick dark stripe along each flank; rich yellow below lateral stripe. Two stump-like forelimbs. Underside white. Two

short, thin rear limbs with two digits on each foot. **DISTRIBUTION** From Balladonia, WA, across southern SA, including the Eyre Peninsula to Gawler. **HABITAT AND HABITS** Found in dune systems, mallee and open woodland. Cathemeral. Fossorial. Shelters under leaf litter, clumps of vegetation, logs and rocks. Eats invertebrates. Lays 1–4 eggs per clutch.

S Eipper

Eastern Robust Slider ■ *Lerista punctatovittata* SVL 11cm

DESCRIPTION Body elongated. Beige to light brown, becoming greyish on to tail, with 4–7 thin black broken stripes that run along body, starting on nape and running on to tail. These are formed by black spot on posterior edge of each dorsal scale. Lateral scales pale yellow with black fleck on anterior edge of each scale. Regenerated tail heavily flecked with black. Underside white. Two stump-like forelimbs, and two short, thin rear limbs

with two digits on each foot. **DISTRIBUTION** Gawler, SA, across to Sea Lake, VIC, through eastern SA, to central NSW, and QLD west of the GDR to Lake Buchanan. **HABITAT AND HABITS** Found in dune systems, mallee, brigalow and open woodland. Cathemeral. Fossorial. Shelters under leaf litter, clumps of vegetation, logs and rocks. Eats invertebrates. Lays 1–6 eggs per clutch.

Dwarf Three-toed Slider ■ *Lerista timida* SVL 5cm

DESCRIPTION Body elongated. Pale to dark brown or grey, usually with fine black flecking. Original tail often yellow with fine brown reticulations. Thick dark stripes along each flank in some populations. Underside bright yellow. Four short, thin limbs with three digits on each foot. Probably a species complex. **DISTRIBUTION** Most of Australia; from Pentland, QLD, south to Stanthorpe, QLD, and west of the GDR through NSW, to the VIC border to near Hattah, through the Flinders Ranges and central SA and the

WA Goldfields. Extends through WA between Port Hedland and Mullewa. **HABITAT AND HABITS** Lives in open woodland, deserts, mallee, brigalow, mulga, grassland and rocky slopes. Cathemeral. Fossorial. Shelters beneath logs, rocks and man-made debris; usually seen disappearing into the soil when uncovered. Eats invertebrates. Lays 1–4 eggs per clutch.

Outcrop Rock Skink ■ *Liburnascincus mundivensis* SVL 6cm

DESCRIPTION Body robust. Head copper coloured. Body silver to greyish-brown with dark grey to black, irregular markings. Regenerated tail plain brown. Usually a dark streak from nasal scale extending to ear. Underside white. Two keels on each scale. Four long limbs with four digits on front feet and five on rear feet. **DISTRIBUTION** Eastern QLD, between Mt Marble and Kings Plain Station west of Cooktown. **HABITAT AND HABITS** Found in rocky outcrops and rocky gorges in open woodland. Diurnal. Shelters beneath logs and rocks. Eats invertebrates. Lays 1–6 eggs per clutch.

S Eipper

Black Mountain Rock Skink ■ *Liburnescincus scirtetis* SVL 7cm

DESCRIPTION Body lightly built. Bronze to metallic greenish-black with fine white to yellow flecking that is most prominent on limbs and lower flanks. Some individuals have grey to yellow vertebral stripe. Four long limbs with four digits on front feet and five on rear feet. **DISTRIBUTION** North-east QLD, at Kalkajaka NP near Rossville. **HABITAT AND HABITS** Found among black granite boulders in tropical open woodland. Diurnal. Very agile. Shelters in rock crevices and under rocks. Eats invertebrates and fruits of native figs. Lays 1–2 eggs per clutch.

S Eipper

Guthega Skink ■ *Liopholis guthega* SVL 10cm

DESCRIPTION Body robust. Pale grey to black. Usually thin, yellow to cream vertebral and dorsolateral stripes. Series of white spots and occelli cover flanks. These markings may loosely align on tail, becoming lighter bands. Underside white to cream. Conspicuous yellow ring around each eye. Markings more prominent in juveniles than adults. Four

muscular limbs with five digits on each foot. **DISTRIBUTION** Restricted to the Australian Alps in two separate populations, one around Mt Kosciuszko, NSW, the other on the Bogong High Plains, VIC. **HABITAT AND HABITS** Found in alpine woodland, heaths and grassland, usually with exposed rocks. Diurnal. Shelters inside rock crevices and under rocks on soil. Mainly eats invertebrates, but occasionally eats fruits and berries. Gives birth to 2–8 live young per litter.

M Sanders/Ecosmart Ecology

Desert Skink ■ *Liopholis inornata* SVL 12cm

DESCRIPTION Body robust. Reddish-brown to cream. Series of black spots that form irregular dorsolateral line. Flanks plain or spotted with pale markings, and flecked with black. Underside white. Four muscular limbs with five digits on front feet and five on rear feet. **DISTRIBUTION** Across southern and central Australia, from south-west QLD, to Jobs Gate, NSW, south to Woomelang, VIC, through SA to Kalgoorlie, WA; extends

to WA coastline between Shark Bay and Port Hedland. **HABITAT AND HABITS** Found in sandy deserts, mallee and hummock grassland. Crepuscular to diurnal. Shelters in elaborate burrows dug into sandy soil beneath vegetation or other structures. Mainly eats vegetation and invertebrates. Gives birth to 1–4 live young per litter.

H Cogger

Great Desert Skink ■ *Liopholis kintorei* SVL 19cm

DESCRIPTION Body robust. Yellowish-brown to orange-red above. Some individuals have brown tipping on dorsal scales forming fine stripes along body. Lower lateral sides yellow or white. Bluish-grey lower lateral zones in some individuals. Underside yellow to white. Four muscular limbs with five digits on front feet and five on rear feet. **DISTRIBUTION** Western central Australia, from Harts Range, NT, south to north-west SA, to Cosmo Newbery, WA, and north to Sturt Creek. **HABITAT AND HABITS** Found in sandy deserts and hummock grassland. Diurnal to crepuscular in warm weather. Shelters in elaborate burrows dug into sandy soil beneath vegetation or other structure. Like some other communal skinks, utilizes shared latrine site. Mainly eats vegetation and invertebrates. Gives birth to 2–6 live young per litter.

S Eipper

Masked Rock Skink ■ *Liopholis margaretae* SVL 12cm

DESCRIPTION Body robust. Pale grey to dark brown, mottled with black markings that form irregular dark stripes and bars. Lower flanks can be flushed with orange. Underside pinkish-orange to cream. Conspicuous yellow to orange ring around each eye. Markings more prominent in juveniles than adults. Four muscular limbs with five digits on each foot. **DISTRIBUTION** Central Australia in the NT, in the east and West Macdonald Ranges. **HABITAT AND HABITS** Found in woodland and rocky gorges. Diurnal. Shelters inside rock crevices and under rocks. Mainly eats invertebrates but occasionally eats vegetation. Gives birth to 2–4 live young per litter.

S Eipper

Eastern Rock Skink ■ *Liopholis modesta* SVL 12cm

DESCRIPTION Body robust. Pale grey to dark brown. Lower flanks can be flushed with orange. Underside pinkish-orange to cream. White markings edged with black along head-sides on to neck, and on sides to shoulder. Conspicuous white ring around each eye. Markings more prominent in juveniles than adults. Four muscular limbs with five digits on each foot. **DISTRIBUTION** Southeast QLD, from Injune south along the GDR, to Singleton, NSW. Isolated population around Yathong NR. **HABITAT AND HABITS** Found in woodland, heaths and rocky grassland. Diurnal. Shelters inside rock crevices and under rocks. Mainly eats invertebrates but occasionally eats vegetation. Gives birth to 2–6 live young per litter.

Adult　　　　　　　　　　　　　　　　*Juvenile*

Night Skink ■ *Liopholis striata* SVL 12cm

DESCRIPTION Body robust. Reddish-brown to orange-red above. Some individuals have brown tipping on dorsal scales forming fine stripes along body. Flanks plain or spotted with pale markings. Lower lateral zone whitish-yellow. Underside yellow to white. Distinctive vertical pupils. Four muscular limbs with five digits on front feet and five on rear feet.

DISTRIBUTION Western central Australia, from Tennant Creek, NT, to north-west SA, to Kalgoorlie, WA, and north to Port Hedland. **HABITAT AND HABITS** Found in sandy deserts and hummock grassland. Crepuscular to nocturnal. Shelters in elaborate burrows dug into sandy soil beneath vegetation or other structures. Like some other communal skinks, utilizes shared latrine site. Mainly eats vegetation and invertebrates. Gives birth to 1–7 live young per litter.

White's Skink ■ *Liopholis whitii* SVL 14cm

DESCRIPTION Body robust. Pale grey to black or brown. Usually pale vertebral and dorsolateral stripes. Series of white spots and occelli cover flanks and in some individuals cover back. These markings may loosely align on tail, becoming lighter bands. Underside yellow to cream. Conspicuous yellow to white ring around each eye. Some individuals almost patternless with a few white spots on sides. Markings more prominent in juveniles than adults. Four muscular limbs with five digits on each foot. **DISTRIBUTION** Eastern Australia, from Mt Barney, QLD, through much of eastern NSW, VIC, except north-west

corner, all of TAS including the Bass Strait Islands, across south south-east SA, to Coffin Bay. Isolated population at Mutawintji NP, NSW. **HABITAT AND HABITS** Found in woodland, heaths and grassland, usually with exposed rocks. Diurnal. Shelters inside rock crevices and under rocks on soil. Mainly eats invertebrates but occasionally eats vegetation. Gives birth to 2–8 live young per litter.

S Eipper

Swamp Skink ■ *Lissolepis coventryi* SVL 11cm

DESCRIPTION Body robust. Pale grey to dark brown mottled with black speckles and fine white to yellow dots. Usually dark vertebral stripe with pair of broad yellowish paravertebral stripes. Upper flanks black; lower flanks grey with pale spots. Conspicuous yellow to white ring around each eye. Underside yellow to white. Markings more prominent in juveniles than adults. Four muscular limbs with five digits on each foot. **DISTRIBUTION** Southeastern Australia, from Millicent, SA, across through southern VIC, to Mallacoota.

HABITAT AND HABITS Found in swamps, creeklines and salt marshes. Diurnal. Usually seen basking on grass tussocks adjacent to shelter site. Shelters inside crab holes, and under man-made debris, rocks and logs. Mainly eats invertebrates but occasionally eats vegetation. Gives birth to 1–6 live young per litter.

A McNab

Mourning Skink ■ *Lissolepis luctuosa* SVL 13cm

DESCRIPTION Body robust. Bronze to dark brown mottled with black markings and fine white to yellow dots. Dark markings form 1–7 irregular stripes that merge on to tail, causing tail to be dark. Conspicuous yellow to white ring around each eye. Upper flanks black; lower flanks heavily spotted with yellow and bronze. Underside yellow with black spots. Markings more prominent in juveniles than adults. Four muscular limbs with five digits on each foot. **DISTRIBUTION** South-west WA, in coastal regions between

Cheyne and Herdsman Lake. **HABITAT AND HABITS** Found in swamps, creeklines and salt marshes. Diurnal; nocturnal in warm weather. Shelters inside crab holes, and under rocks, logs and man-made debris. Usually seen basking near cover. Mainly eats invertebrates but occasionally eats vegetation. Gives birth to 2–5 live young per litter.

Tree-base Litter-skink ■ *Lygisaurus foliorum* SVL 4cm

DESCRIPTION Body elongated. Pale brown to dark brown with pale flecks. Throats on breeding males can be flushed with orange. Lips barred with black. Underside cream to white. Four thin limbs with four digits on front feet and five on rear feet.

DISTRIBUTION Between the Lynd, QLD, over the GDR and Campbelltown, NSW. Extends as far inland as Torrens Creek, QLD, in north, to the Warrumbungles, NSW, in south. **HABITAT AND HABITS** Found in open woodland, brigalow and grassland. Diurnal. Fossorial. Shelters in leaf litter, and under fallen timber and rocks. Mainly eats invertebrates. Lays two eggs.

Orange-speckled Forest Skink ■ *Magmellia lutillateralis* SVL 10cm

DESCRIPTION Body robust. Dark brown to dark grey with fine irregular black bars, becoming darker on tail. Flanks blackish-grey with fine white flecks; lower flanks bright orange. Underside pale yellow. Four muscular limbs with five digits on each foot. **DISTRIBUTION** Above 900m on the Eungella Plateau, north-east QLD. **HABITAT AND HABITS** Found in closed forest and rainforest. Diurnal. Shelters in rotten logs, and under fallen timber and rocks. Usually seen basking on edges of logs along walking trails in forests. Mainly eats invertebrates. Gives birth to live young.

S Eipper

Common Dwarf Skink ■ *Menetia greyii* SVL 4cm

DESCRIPTION Body lightly built and elongated. Grey to brown, with or without grey to black flecking. Usually has broad, dark-coloured lateral stripe bordered by white stripe below. Throat in breeding males can have orange flush. Underside white. Four limbs with four digits on front feet and five on rear ones. **DISTRIBUTION** Across Australia, west of the GDR. Does not occur in north-east NT, on Cape York Peninsula, QLD, southern VIC or TAS. **HABITAT AND HABITS** Found in deserts, heaths, grassland, woodland and urban environments. Diurnal. Fossorial. Shelters beneath logs and rocks. Most commonly seen while moving through leaf litter. Eats invertebrates. Lays 1–3 eggs per clutch.

S Eipper

Saltbush Morethia ■ *Morethia adelaidensis* SVL 6cm

DESCRIPTION Body medium in build. Grey to brown. Back predominantly plain with irregular black blotches that can form broken stripes. Flanks pale grey to cream with irregular dark blotches. Throat in breeding males can have orange-red flush. Underside white. Four limbs with five digits on each foot.

DISTRIBUTION Across the Eyrean Basin and southern Australia, between Kalgoorlie, WA, and Kerang, VIC. **HABITAT AND HABITS** Found in deserts, heaths, grassland, woodland and urban environments. Diurnal. Shelters beneath logs and rocks. Most commonly seen while moving through leaf litter. Eats invertebrates. Lays 1–3 eggs per clutch.

Boulenger's Morethia ■ *Morethia boulengeri* SVL 5cm

DESCRIPTION Body medium in build. Grey to brown. Back predominantly plain with irregular black flecks. Flanks have broad black stripe from nostril on to tail; stripes bordered below with white. Tail orange in young individuals. Underside white. Throat in breeding males can have orange-red flush. Four limbs with five digits on each foot. **DISTRIBUTION**

West of the GDR, between Moranbah, QLD, through NSW, into Victoria as far south as Seymour, west through SA and southern NT, to Illkulka, WA. **HABITAT AND HABITS** Found in deserts, heaths, grassland, woodland and urban environments. Diurnal. Shelters beneath logs and rocks. Most commonly seen while basking. Eats invertebrates. Lays 1–3 eggs per clutch.

Lined Fire-tailed Skink ■ *Morethia ruficauda* SVL 4.5cm

DESCRIPTION Body medium in build. Body and flanks black with thin yellow dorsolateral and lateral stripes that extend from snout to hips. Hips, including rear legs and tail, bright reddish-orange. Underside white. Four limbs with five digits on each foot. M. r. *exquisita* has additional yellow vertebral stripe.
DISTRIBUTION M. r. *ruficauda* from far western QLD, through the NT and the Kimberley, WA; M. r. *exquisita* through the Pilbara into the Gascoyne drainage. **HABITAT AND HABITS** Found in deserts, grassland, woodland and rocky gorges. Diurnal. Shelters beneath logs and rocks. Most commonly seen while basking. Eats invertebrates. Lays 1–3 eggs per clutch.

A Zimny

Nangur Skink ■ *Nangura spinosa* SVL 10cm

DESCRIPTION Body robust and rough scaled. Pale brown with series of dark, thin irregular bands. Whole upper body flecked with pale brown. Underside pale with brown spotting. Four robust limbs with five digits on each foot. **DISTRIBUTION** Nangur and Oakview NP, south-east QLD. **HABITAT AND HABITS** Found in vine thickets and rainforest. Diurnal. Shelters in burrows constructed beneath fallen timber, tree roots and rocks. Sits at entrance of burrow on bare patch of soil. Mainly eats invertebrates. Gives birth to live young that live in burrow with female for at least nine months.

S Eipper

Ornate Snake-eyed Skink ■ *Notoscincus ornatus* SVL 4cm

DESCRIPTION Body medium in build. Copper to brown. Back predominantly plain with irregular black flecks. Flanks have broad black stripe with reddish-orange flecks evenly spaced along its length; stripe occurs from nasal scale, along body, to tail-tip. Lateral stripes

bordered below with white. Eyes large. Underside white. Four limbs with five digits on each foot. **DISTRIBUTION** From the North West Cape, across through the Kimberley, across the NT to the SA border, through western and northern QLD to Charters Towers. **HABITAT AND HABITS** Found in deserts, grassland and woodland. Diurnal. Shelters beneath logs and rocks. Eats invertebrates. Lays 1–4 eggs per clutch.

Yolk-bellied Snake-skink ■ *Ophioscincus ophioscincus* SVL 7cm

DESCRIPTION Body elongated. Beige to light grey with black flecking. Flanks black.

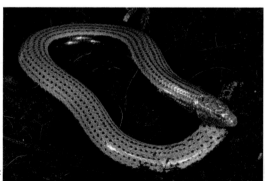

Underside bright yellow. Completely limbless. **DISTRIBUTION** Between Brisbane and Miriam Vale, south-east QLD. **HABITAT AND HABITS** Lives in woodland, rainforest, rocky slopes and moist gardens. Nocturnal. Fossorial. Shelters beneath logs, rocks and man-made debris. Usually seen disappearing into the soil when uncovered. Eats invertebrates. Egg laying.

Tussock Skink ■ *Pseudemoia pagenstecheri* SVL 6cm

DESCRIPTION Body medium in build. Bronze-brown to greyish-silver with series of prominent black stripes. Pale dark dorsolateral stripe centred on fourth scale row from midline. White to cream lower lateral stripe extends from sides of ear to tail. Stripe becomes reddish-orange in breeding males. Underside white. Four limbs with five digits on each foot. **DISTRIBUTION** From Lucinda, SA, east through southern VIC, and across the Southern Highlands in NSW, up to Guyra. Also TAS. **HABITAT AND HABITS** Occupies wet forests, open woodland, grassland and urban environments. Diurnal. Shelters beneath logs and rocks. Eats invertebrates. Gives birth to 2–7 live young per litter.

S Eipper

Glossy Grass Skink ■ *Pseudemoia rawlinsoni* SVL 6cm

DESCRIPTION Body medium in build. Bronze-brown to greyish-silver with series of prominent black stripes. Pale dark dorsolateral stripe centred on third scale row from midline. White to cream lower lateral stripe extends from sides of ear to tail. Stripe becomes reddish-orange in breeding males. Underside white. Four limbs with five digits on each foot. **DISTRIBUTION** From Robe, SA, east through southern VIC, and across the Southern Highlands in NSW and the ACT, to Coree. Also north-east TAS and Cape Barren Island. **HABITAT AND HABITS** Occupies wet forests, open woodland, grassland and swamps. Diurnal. Shelters beneath logs and rocks. Eats invertebrates. Gives birth to 2–6 live young per litter.

S Eipper

Spencer's Skink ■ *Pseudemoia spenceri* SVL 7cm

DESCRIPTION Body medium in build. Dark grey to brown with black and coppery-brown flecks, and pair of yellow dorsolateral stripes that extend on to tail. White to yellow lower lateral stripes. Lower flanks dark mottled grey to black. Underside cream. Four limbs with five digits on each foot. **DISTRIBUTION** From the Grampians in southern VIC, and across into the Southern Highlands in NSW; disjunct population in the Blue Mountains, NSW. **HABITAT AND HABITS** Lives in wet forests and open woodland. Diurnal. Shelters beneath logs and rocks. Lives in groups of up to 30 individuals in rock crevices. Eats invertebrates. Gives birth to 2–6 live young per litter.

Three-toed Skink ■ *Saiphos equalis* SVL 9cm

DESCRIPTION Body elongated. Dark brown or grey, usually with black flecking. Flanks black. Underside bright yellow. Four short, thin limbs with three digits on each foot. **DISTRIBUTION** Wagga Wagga, NSW, along east coast and GDR to Nangur NP, QLD. Isolated populations at Kroombit Tops and Carnarvon Gorge NP. **HABITAT AND**

HABITS Lives in woodland, rainforest, rocky slopes and moist gardens. Nocturnal. Fossorial. Shelters beneath logs, rocks and man-made debris. Usually seen disappearing into the soil when uncovered. Commonly mistaken for a small snake. Eats invertebrates. Bimodal reproduction, with coastal populations laying 1–5 thin-shelled eggs and mountain populations being live bearing.

Southern Wet Tropics Shadeskink ■ *Saproscincus basiliscus* SVL 5cm

DESCRIPTION Body medium in build. Reddish-brown to dark brown, with pale and dark flecking. Rusty dorsolateral stripe extends from snout on to tail. Dark stripe from nasal scale over shoulder to midway along flanks. Underside cream to white. Four limbs with five digits on each foot.

DISTRIBUTION Between Bloomfield and Mt Elliot, QLD. Isolated population at Conway NP, QLD.

HABITAT AND HABITS Found in rainforest, closed woodland and gardens. Diurnal. Shelters in leaf litter, and under fallen timber and rocks; also found at night sleeping on leaves. Mainly eats invertebrates. Egg laying.

S Eipper

Eungella Shadeskink ■ *Saproscincus eungellensis* SVL 7cm

DESCRIPTION Longest of the shadeskinks. Body medium in build and elongated. Pale brown to dark brown with pale and dark flecking. Rusty dorsolateral stripe extends from hips on to tail. Dark stripe from nasal scale over shoulder. Head-sides pale from snout to ears. Underside cream to white with dark streaks and flecking. Four limbs with five digits on each foot.

DISTRIBUTION Clarke Range above 700m, near Eungella, QLD. **HABITAT AND HABITS** Found in rainforest, particularly in gullies, soaks and creeklines. Diurnal. Seen at night sleeping on fern tips. Shelters in leaf litter, and under fallen timber and rocks. Mainly eats invertebrates. Egg laying.

S Eipper

Weasel Skink ■ *Saproscincus mustelina* SVL 6cm

DESCRIPTION Body medium in build. Reddish-brown to dark brown, with pale and dark flecking. Yellow to rusty dorsolateral stripe extends from midbody on to tail. Conspicuous

white spot behind eye. Underside cream to white with brown flecking. Original tail very long. Four limbs with five digits on each foot. **DISTRIBUTION** Between Guyra, NSW, along the GDR, to the Otway Ranges, VIC. **HABITAT AND HABITS** Found in rainforest, closed woodland and gardens. Diurnal. Shelters in leaf litter, and under fallen timber and rocks. Mainly eats invertebrates. Lays 1–6 eggs in clutch.

S Eipper

Rose's Shadeskink ■ *Saproscincus rosei* SVL 5cm

DESCRIPTION Body medium in build. Reddish-brown to dark brown, with pale and dark flecking. Rusty dorsolateral stripe extends from shoulder on to tail. Females can have broad black lateral stripe with prominent yellow lower stripe below. Underside cream to

white. Four limbs with five digits on each foot. **DISTRIBUTION** Between Oakview NP, QLD, along the GDR, to the Barrington Tops, NSW. **HABITAT AND HABITS** Found in rainforest, closed woodland and gardens. Diurnal. Shelters in leaf litter, and under fallen timber and rocks, but can be found at night sleeping on leaves. Mainly eats invertebrates. Lays 2–5 eggs in a clutch.

S Eipper

Four-fingered Shadeskink ■ *Saproscincus tetradactylus* SVL 3cm

DESCRIPTION Body medium in build. Reddish-brown to dark brown, with fine black longitudinal streaks. Strongly pointed snout. Lips barred with white. Rusty dorsolateral stripe over hips extending on to tail. Underside cream to white. Very shiny scales refract in sunlight, giving rainbow-like appearance. Four limbs with four digits on front feet and five on rear feet.
DISTRIBUTION Between Paluma and Kuranda, QLD.
HABITAT AND HABITS Found in rainforest. Diurnal. Fossorial. Shelters in leaf litter, and under fallen timber and rocks. Mainly eats invertebrates. Egg laying.

S Eipper

Pygmy Blue-tongue ■ *Tiliqua adelaidensis* SVL 10cm

DESCRIPTION Body robust. Yellowish-brown, grey brown to dark brown, with fine dark streaks running along body. These black markings can coalesce to form black irregular blotches. Lower lateral zone cream, pale yellow to light grey. Underside off white, sometimes with fine grey streaks. Tongue pale pink and mouth lining mauve. Four muscular limbs with five digits on each foot. **DISTRIBUTION** From Peterborough to Kapunda, SA. Historically occurred in Adelaide's outer suburbs. **HABITAT AND HABITS** Found in treeless grassland and heavy reddish clay soils. Diurnal. Once thought to be extinct, and rediscovered from an individual found inside a road-killed brown snake. Shelters in spider burrows and has been found under stones. Mainly eats invertebrates. Gives birth to 1–4 young per litter.

J Meney

Centralian Blue-tongue ■ *Tiliqua multifasciata* SVL 25cm

DESCRIPTION Body robust. Creamy-white to pale grey with 16–25 yellow to reddish-orange, broad bands. Prominent black temporal stripe extends from rear of eye to above ear. Underside off white, sometimes with fine darker streaks formed by dark margins on edges of ventral and gular scales. Tongue bright blue and mouth lining pink. Four muscular

limbs with five digits on each foot. **DISTRIBUTION** From Longreach, west through QLD, across the NT and far northern SA, to the Pilbara in WA. **HABITAT AND HABITS** Found in deserts, grassland and open woodland; also on rocky slopes with spinifex. Diurnal, but crepuscular in warm weather. Shelters beneath vegetation, logs and rocks. Eats vegetation and invertebrates. Gives birth to 2–9 live young per litter.

Blotched Blue-tongue ■ *Tiliqua nigrolutea* SVL 32cm

DESCRIPTION Body robust. Pale brown to dark brown, grey or black, with prominent white, yellow, orange, red to pale brown irregular blotches. Underside pale brown, orange or yellow, with irregular dark grey to black barring. Tongue bright blue and mouth lining pink. Four muscular limbs with five digits on each foot. **DISTRIBUTION** NSW from the

Newnes Plateau, along the GDR, south into southern VIC, and across to Robe in SA. Also across TAS. **HABITAT AND HABITS** Found in grassland, heaths, swamps and woodland; common in urban environments. Diurnal, but crepuscular in warm weather. Shelters beneath vegetation, logs, rocks and man-made debris. Eats vegetation and invertebrates. Gives birth to 1–9 live young per litter.

Western Blue-tongue ■ *Tiliqua occipitalis* SVL 32cm

DESCRIPTION Body robust. Creamy-white, yellow to pale grey, with 8–13 dark grey to chocolate-coloured bands. Prominent black stripe extends from snout to above ear. Underside off white to pale yellow, sometimes with fine dark streaks or dark blotches. Tongue dark blue and mouth lining pink. Four muscular limbs with five digits on each foot.

DISTRIBUTION From Round Hill, NSW, through western NSW and north-west VIC, through SA into the NT as far north as Yulara, and across to Carnarvon in WA. **HABITAT AND HABITS** Found in deserts, grassland, mallee, chenopod shrubland and open woodland. Diurnal, but crepuscular in warm weather. Shelters beneath vegetation, logs and rocks. Eats vegetation and invertebrates. Gives birth to 2–13 live young per litter.

S Eipper

Shingleback Lizard ■ *Tiliqua rugosa* SVL 32cm

DESCRIPTION Body robust. Colouration and pattern very variable; may be any combination of white, yellow, orange, red, brown, grey and black; some individuals banded, some flecked and others plain. Tongue dark blue and mouth lining pink. Four muscular limbs with five digits on each foot. Four subspecies. **DISTRIBUTION** *T. r. rugosa* from Balladonia across south-west WA, to Gee Gie Outcamp; *T. r. aspera* from Bunningonia Springs, WA, west through SA, into QLD south of Aramac, NSW, ACT and VIC, west of the GDR; *T. r. konowi* restricted to Rottnest Island, WA; *T. r. palarra* between Carnarvon and Tamala, WA. **HABITAT AND HABITS** Found in deserts, grassland, mallee, chenopod scrubland and open woodland. Diurnal. Shelters beneath vegetation, logs and rocks. Eats vegetation, invertebrates and food scraps. Gives birth to 1–3 live young per litter.

S Eipper

Common Blue-tongue ■ *Tiliqua scincoides* SVL 49cm

DESCRIPTION Body robust. Highly variable; pale brown to dark brown, grey or black, with prominent yellow, orange-brown to black irregular bands. Individuals from the Kimberley

can be heavily flecked with bluish-grey markings. Usually dark temporal stripe. Underside white, pale pink, orange or yellow, with irregular dark grey to black barring. Tongue bright blue and mouth lining pink. Four muscular limbs with five digits on each foot. Two poorly defined subspecies. **DISTRIBUTION** *T. s. scincoides* from Cooktown, through QLD, NSW and VIC, across into eastern SA. Isolated population in northern SA and southern NT. *T. s. intermedia* on Cape York Peninsula, through the Gulf region of QLD, as far south as Richmond, QLD, and across through the NT, north of Tennant Creek, to Broome, WA. **HABITAT AND HABITS** True generalist, found in deserts, woodland, heaths and grassland; common in urban environments. Diurnal, but can be crepuscular in warm weather. Shelters beneath vegetation, logs, rocks and man-made debris. Eats vegetation, invertebrates, food scraps and carrion. Gives birth to 2–42 live young per litter.

T. s. scincoides

T. s. intermedia

Dragons

Burn's Dragon ■ *Amphibolurus burnsi* SVL 12cm

DESCRIPTION Body medium in build. Grey to brown with cream to yellow stripe extending from snout along flanks to hips. Short spines run along vertebral and nuchal ridge, forming low crests. **DISTRIBUTION** QLD, SA and NSW, between Llanarth, QLD, Innamincka, SA, and Nymagee, NSW. **HABITAT AND HABITS** Found in dry

open woodland, brigalow, chenopod shrubland and mallee. Diurnal. Often seen by day perched on trees and fence posts. Communicates with other dragons with head bobbing, press-ups and arm waves, hence the alternative name Ta-Ta Lizard. Feeds on invertebrates. At night may be seen sleeping on branches. Lays 3–11 eggs per clutch.

Jacky Dragon ■ *Amphibolurus muricatus* SVL 13cm

DESCRIPTION Body medium in build. Pale grey to brown, usually with series of paler broad blotches coalescing to form band along either side of spine from shoulder on to hips. Breeding males can be black with white stripe extending from snout along flanks. Can alter intensity of colour and pattern, depending on temperature and reproductive status. Short spines run along vertebral and nuchal ridge, forming low crests. Probably a species complex. **DISTRIBUTION** From Girraween, QLD, through eastern NSW, as far west as the Warrumbungles into southwestern VIC. **HABITAT AND HABITS** Found in open woodland, heaths, grassland verges and closed forest. Diurnal. Often seen by day perched on trees and fence posts. Feeds on invertebrates. Lays 2–14 eggs per clutch.

S Eipper

Mulga Dragon ■ *Caiminops amphiboluroides* SVL 10cm

DESCRIPTION Body lightly built. Pale grey to brown, usually with series of paler broad blotches coalescing to form band along either side of spine from shoulder on to hips. Breeding males can be black with white stripe extending from snout along flanks. Can alter intensity of colour and pattern, depending on temperature and reproductive status. Long spines run along vertebral and nuchal ridge, forming low crests. Sometimes placed in genus *Diporiphora*. **DISTRIBUTION** From WA Goldfield and Pilbara regions between Gahnda Rockhole, north to Karijini and as far south as Kanowna. **HABITAT AND HABITS** Found in open woodland, mulga and mallee. Rarely seen due to excellent camouflage. Diurnal. Feeds on invertebrates. Lays 2–7 eggs per clutch.

A Elliott

Chameleon Dragon ■ *Chelosania brunnea* SVL 12cm

DESCRIPTION Body robust. Vertically compressed with distinctive dewlap. Low nuchal crest. Pale grey to reddish-brown, usually with broad dark grey bands on tail. Breeding

males can have yellow flush on lower jaw on to throat. **DISTRIBUTION** From Doomadgee, QLD, through Top End region of the NT, to Dampier, WA. **HABITAT AND HABITS** Found in tropical open woodland. Diurnal. Rarely seen but occasionally found perched in trees and shrubs. Poorly known due to secretive nature. Feeds on invertebrates. Lays 5–9 eggs per clutch.

M Sanders/Ecosmart Ecology

Frilled Dragon ■ *Chlamydosaurus kingii* SVL 30cm

DESCRIPTION Body medium in build. Pale grey to black and every shade of brown. Can alter intensity of colour and pattern, depending on temperature and reproductive status. Short spines run along vertebral ridge, forming low crest. Frill can have black spots, usually red, orange, yellow or white on lower third of frill; frill colouration linked to distribution. **DISTRIBUTION** Tropical Australia, from Greenbank, QLD, to Broome, WA. **HABITAT**

AND HABITS Found in dry woodland, rocky gorges and tropical savannah. Diurnal. Often seen by day perched on trees or on the ground adjacent to a tree. When threatened, usually flees up a tree. If caught out in the open can run bipedally to escape a predator. Its last line of defence is the frill, which is erected, usually with the mouth open, to scare off a predator. Feeds on invertebrates. At night can sometimes be seen sleeping on branches. Lays 3–14 eggs per clutch.

S Eipper

Crested Bicycle Dragon ■ *Ctenophorus cristatus* SVL 12cm

DESCRIPTION Body medium in build. Pale grey to brown, usually with broken dark stripe along either side of body, extending from snout to hips, coupled with black flecks and spots randomly over forebody. Tail broadly banded. Male often flushed with bright orange or yellow. Can alter intensity of colour and pattern, depending on temperature and reproductive status. Short spines run along vertebral ridge, forming low crest. **DISTRIBUTION** Southern Australia, from the Eyre Peninsula, SA, to the Wongan Hills,

WA. **HABITAT AND HABITS** Found in dry open woodland and mallee shrubland. Diurnal. Often seen by day perched on trees and fallen timber. Feeds on invertebrates but also eats small skinks. At night sometimes seen sleeping inside hollow logs. Lays 2–9 eggs per clutch.

S Scott

Tawny Dragon ■ *Ctenophorus decressi* SVL 11cm

DESCRIPTION Body medium in build. Pale to dark grey or brown, usually with broken dark stripe along either side of body, extending from snout to hips. Male often flushed with bright orange or yellow. Can alter intensity of colour and pattern, depending on temperature and reproductive status. Probably a species complex. **DISTRIBUTION**

Southern Australia, from Kangaroo Island to northern Flinders Ranges, SA. **HABITAT AND HABITS** Found in dry open woodland, rocky gorges and mallee shrubland. Diurnal. Often seen by day perched on rocks and boulders. Feeds on invertebrates. Lays 3–7 eggs per clutch.

S Eipper

Peninsula Dragon ■ *Ctenophorus fionni* SVL 10cm

DESCRIPTION Body medium in build. Female white to dark grey or brown, usually with broken light blotches across body. Male dark grey to black, with significant variation in light-coloured patches. These patches, spots, streaks and transverse bands can be bright red, orange, yellow or white. Can alter intensity of colour and pattern, depending on temperature and reproductive status. **DISTRIBUTION** Southern Australia, from the Eyre Peninsula to Lake Torrens, SA. **HABITAT AND HABITS** Found in dry open woodland, heaths, rocky gorges and mallee shrubland. Diurnal. Often seen by day perched on rocks and boulders. If disturbed, flees under rocks or retreats back into crevice. Feeds on invertebrates. Lays 2–7 eggs per clutch.

Variations of species

Mallee Military Dragon ■ *Ctenophorus fordi* SVL 6cm

DESCRIPTION Body medium in build. Pale yellow to reddish-brown, with white to yellow dorsolateral stripe edged with black along either side of body extending from shoulder over hips, coupled with black flecks and spots randomly over forebody. Tail same colour as back, with black and white lateral stripes. Pattern stronger in male than female. Underside white in both sexes. Probably a species complex. **DISTRIBUTION** Southern Australia, from the Goldfields region, over WA, through SA, including the Eyre Peninsula and north-west VIC. **HABITAT AND HABITS** Found in mallee shrubland. Diurnal. Often seen by day running between clumps of spinifex. Feeds on ants and occasionally other small insects. Lays 2–3 eggs per clutch.

Gibber Dragon ■ *Ctenophorus gibba* SVL 13cm

DESCRIPTION Body robust. Pale beige to yellowish-brown with dark and light mottling and flecking. Breeding male has black line from throat on to chest. Underside white in both sexes. **DISTRIBUTION** SA, from Coober Pedy, north to Witjira NP, and east to Marree. **HABITAT AND HABITS** Found on rocky soils with arid grassland, gibber desert and open mulga woodland. Diurnal. Often seen by day perched on rocks and boulders. If disturbed, flees into a burrow, usually under rock it is perched on. Feeds on invertebrates. Lays 7–9 eggs per clutch.

Central Military Dragon ■ *Ctenophorus isolepis* SVL 7cm

DESCRIPTION Body medium in build. Pale yellow to reddish-brown with thin white to yellow dorsolateral stripe edged with black along either side of body, extending from shoulder over hips, coupled with black flecks and spots randomly over forebody. Tail same colour as back, with black and white lateral stripes. Pattern stronger in male than female; male often black laterally with bright yellow flecks. Underside white in both sexes. Probably a species complex. **DISTRIBUTION** Western and central Australia, from the Goldfields region north to southern edge of the Kimberley, across to western QLD and north-east SA. **HABITAT AND HABITS** Found in mallee shrubland, sand-ridge deserts and semi-arid grassland. Diurnal. Often seen by day running between clumps of spinifex. Feeds on ants and occasionally other small insects. Lays 1–6 eggs per clutch.

Male   *Female*

Lake Eyre Dragon ■ *Ctenophorus maculosus* SVL 9cm

DESCRIPTION Body medium in build. Pale beige to whitish-grey with darker mottling and flecks. Usually a series of black spots forming paravertebral row. Breeding male can have orange flushing on forebody and head. Underside white in both sexes. **DISTRIBUTION** SA, between Lake Eyre North, Lake Torrens and Lake Frome. **HABITAT AND HABITS** Found on dry salt crust of evaporated lakes and surrounding open mulga woodland. Diurnal. Often seen by day perched on salt crust. If disturbed, flees into burrow beneath salt crust. Feeds on invertebrates. Lays 2–4 eggs per clutch.

Central Netted Dragon ■ *Ctenophorus nuchalis* SVL 12cm

DESCRIPTION Body robust. Female reddish-brown with broken light blotches forming pair of broken paravertebral stripes; fine white spots cover rest of body. Male blackish-brown with small red to orange spots that are brightest on head and along spine. Underside white in both sexes. **DISTRIBUTION** Central Australia, from Bollon, QLD, west to

WA coastline between Columb Point NR, south to Arrowsmith. East through WA and SA, to Mungo NP, NSW. **HABITAT AND HABITS** Found on rocky soils with arid grassland and open mulga woodland. Diurnal. Often seen by day perched on rocks and boulders. If disturbed flees into burrow, usually under rock it is perched on. Feeds on invertebrates. Lays 3–15 eggs per clutch.

Ornate Dragon ■ *Ctenophorus ornatus* SVL 10cm

DESCRIPTION Body strongly depressed. Female brown, usually with broken light blotches centred along midline, with fine white spots forming transverse bands on flanks and poorly defined tail-rings. Male dark grey to black with broken light blotches centred along midline; fine white spots forming transverse bands on flanks, and stark black and white, alternating rings on tail. Underside grey with reddish markings on forebody. Inland form brown with black and white-ringed tail and white vertebral zone. Probably a species complex. **DISTRIBUTION** South-west WA, from Murchison region to Cape Le Grand. **HABITAT AND HABITS** Found in open woodland, heaths and mallee shrubland. Diurnal. Often seen by day perched on rocks and boulders. If disturbed, flees under rocks or retreats into its crevice. Feeds on invertebrates. Lays 2–6 eggs per clutch.

Painted Dragon ■ *Ctenophorus pictus* SVL 11cm

DESCRIPTION Body robust. Female and juveniles pale whitish to dark grey or brown, generally with broken dark stripe along either side of body extending from snout to hips. Male variable, from bright yellow and blue, to red and blue, usually with dark patterns and white flecks. Can alter intensity of colour and pattern, depending on temperature and reproductive status. Underside white in both sexes. Probably a species complex. **DISTRIBUTION** Arid parts of southern Australia, from north-west VIC, to Thargomindah, QLD, across to Kalgoorlie, WA. **HABITAT AND HABITS** Found in mallee shrubland and mulga woodland. Diurnal. Often seen by day running between clumps of spinifex. Feeds on invertebrates. Lays 5–8 eggs per clutch.

Western Netted Dragon ■ *Ctenophorus reticulatus* SVL 10cm

DESCRIPTION Body robust. Female reddish-brown with broken light blotches forming pair of broken paravertebral stripes, and fine white spots covering remainder of body. Male blackish-brown with small, red to orange spots that are brightest on head and along spine. Underside white in both sexes. **DISTRIBUTION** Southern two-thirds of WA, from Shark Bay across to Eridunda, NT, and south to Lake Everard, SA. **HABITAT AND HABITS** Found on rocky soils with arid grassland and open mulga woodland. Diurnal. Often seen by day perched on rocks and boulders. If disturbed, flees into a burrow, usually under rock it is perched on. Feeds on invertebrates. Lays 5–13 eggs per clutch.

S Eipper

Slater's Ring-tailed Dragon ■ *Ctenophorus slateri* SVL 10cm

DESCRIPTION Body medium in build. Female reddish-brown with broken light blotches and fine white spots forming transverse bands; poorly defined or no tail-rings. Male reddish-brown with or without fine and black white spots forming transverse bands; stark

S Eipper

black and white alternating rings. **DISTRIBUTION** From the Kimberley, WA, across to Longreach in QLD, including rocky regions of most of the NT. **HABITAT AND HABITS** Found on rocky outcrops in arid grassland, open woodland and rocky gorges. Diurnal. Often seen by day perched on rocks and boulders. If disturbed, flees under rocks or retreats into crevices. Feeds on invertebrates. Lays 5–11 eggs per clutch.

Claypan Dragon ▪ *Ctenophorus salinarum* SVL 11cm

DESCRIPTION Body robust. Usually pale white to grey, with series of black, white and pale yellow blotches forming mottled appearance. Male often has bright yellow to orange spots with black throat. Can alter intensity of colour and pattern, depending on temperature and reproductive status. Underside white in both sexes. **DISTRIBUTION** Arid parts of southern Australia, from western SA, through the arid Goldfields region of WA. **HABITAT AND HABITS** Found in mallee shrubland and mulga woodland. Diurnal. Often seen by day running between vegetation. Feeds on invertebrates. Lays 5–8 eggs per clutch.

Lozenge-marked Dragon ▪ *Ctenophorus scutellatus* SVL 12cm

DESCRIPTION Body medium in build. Pale grey to brown, usually with series of broken dark stripes along either side of body extending from snout to hips, coupled with black flecks and white spots randomly over body. Male often flushed with red, orange or yellow. Can alter intensity of colour and pattern, depending on temperature and reproductive status. Short spines run along vertebral ridge, forming low crest. **DISTRIBUTION** WA, between Warburton, Lyndon and Bruce Rock. **HABITAT AND HABITS** Found in dry open woodland and mallee shrubland. Diurnal. Often seen by day perched on trees and fallen timber. Feeds on invertebrates and also small skinks. Lays 4–10 eggs per clutch.

Tommy Roundhead ■ *Diporiphora australis* SVL 7cm

DESCRIPTION Body lightly built. Pale grey to yellow or brown, with or without series of broad blotches within broad pale band along either side of spine from shoulder on to hips. Can alter intensity of colour and pattern, depending on temperature and reproductive

status. Strong gular fold. **DISTRIBUTION** From Grafton, NSW, up along eastern coast to Normanton, and across Cape York to Cooktown, QLD. **HABITAT AND HABITS** Found in open woodland, heaths, grassland and dry forest. Diurnal. Often seen by day perched on trees and fence posts. Communicates with other dragons with head bobbing, press-ups and arm waves. Feeds on invertebrates. Lays 2–6 eggs per clutch.

Granulated Dragon ■ *Diporiphora granulifera* SVL 8cm

DESCRIPTION Body lightly built. Pale grey to yellow or brown, with or without series of broad blotches within broad pale band along either side of spine from shoulder on to hips. Large males can be plain olive-green to grey with yellow lateral stripe with pink flush on tail-base. Can alter intensity of colour and pattern, depending on temperature and

reproductive status. Strong gular fold. **DISTRIBUTION** North-west QLD, along the Selwyn Range from Winton to Doomadgee. **HABITAT AND HABITS** Found in open woodland, tropical grassland and dry forest. Diurnal. Often seen by day perched on trees and running between clumps of vegetation. Communicates with other dragons with head bobbing, press-ups and arm waves. Feeds on invertebrates. Lays 2–6 eggs per clutch.

Cape York Two-lined Dragon ■ *Diporiphora jugularis* SVL 7cm

DESCRIPTION Body lightly built. Yellow to brown with or without series of broad blotches within broad pale band along either side of spine from shoulder on to hips. Breeding males can have striking black throat. Can alter intensity of colour and pattern, depending on temperature and reproductive status. No gular fold. **DISTRIBUTION** From Cooktown, through northeastern QLD, between Koah and Mer Island in the Torres Strait. **HABITAT AND HABITS** Found in tropical open woodland and heaths. Diurnal. Often seen by day perched on trees and fence posts. Communicates with other dragons with head bobbing, press-ups and arm waves. Feeds on invertebrates. Lays 2–6 eggs per clutch.

Male in breeding colours   *Non-breeding colours*

Lally's Two-lined Dragon ■ *Diporiphora lalliae* SVL 7cm

DESCRIPTION Body lightly built. Yellow to brown with series of broad orange to black blotches within broad pale yellow band along either side of spine from shoulder on to hips. Can alter intensity of colour and pattern, depending on temperature and reproductive status. Strong gular fold.
DISTRIBUTION From Eighty Mile Beach, WA, across through central NT, to Camooweal, QLD.
HABITAT AND HABITS Found in sand-ridge deserts, semi-arid grassland and mallee shrubland. Diurnal. Often seen sitting on dead trees and termite mounds. Feeds on small invertebrates. Lays 1–6 eggs per clutch.

Yellow-sided Two-lined Dragon ■ *Diporiphora magna* SVL 7cm

DESCRIPTION Body lightly built. Green to yellowish-brown with or without series of broad blotches within broad pale band along either side of spine from shoulder on to hips. Breeding males can have large black spot on shoulder. Underside yellow to white. Can alter intensity of colour and pattern, depending on temperature and reproductive

status. No gular fold. **DISTRIBUTION** From Fitzroy Crossing, WA, through northern NT, to Calvert. **HABITAT AND HABITS** Found in tropical open woodland. Diurnal. Often seen by day perched on trees and fence posts. Communicates with other dragons with head bobbing, press-ups and arm waves. Feeds on invertebrates. Lays 2–6 eggs per clutch.

Common Nobbi Dragon ■ *Diporiphora nobbi* SVL 8cm

DESCRIPTION Body lightly built. Yellow to brown with or without series of paler broad blotches coalescing to form band along either side of spine from shoulder on to hips. Breeding males can have pink flush on hips and tail. Can alter intensity of colour and pattern, depending on temperature and reproductive status. Short spines run along vertebral and nuchal ridge, forming low crests. Probably a species complex.

DISTRIBUTION From Cooktown, through eastern QLD and NSW, into northwestern VIC and eastern SA. **HABITAT AND HABITS** Found in open woodland, heaths, grassland verges and mallee. Often seen by day perched on trees and fence posts. Communicates with other dragons with head bobbing, press-ups and arm waves. Feeds on invertebrates. Lays 2–14 eggs per clutch.

Superb Dragon ■ *Diporiphora superba* SVL 10cm

DESCRIPTION Body lightly built and elongated. Tail very long. Tan to vivid green, with or without white-edged, broad maroon vertebral stripe. Underside yellow. Juveniles initially pale brown.

DISTRIBUTION North-west Kimberley region, WA.
HABITAT AND HABITS Found in tropical woodland and woodland bordering sandstone gorges and rocky creeks. Diurnal. Strongly arboreal; usually seen in shrubs and trees. Moves slowly, resembling wind moving through branches. Also seen sleeping on edges of vegetation at night. Feeds on invertebrates. Lays five eggs per clutch.

Canegrass Dragon ■ *Diporiphora winneckei* SVL 7cm

DESCRIPTION Body lightly built. Pale grey to yellow or brown with series of broad blotches within broad pale band along either side of spine from shoulder on to hips. Pair of pale yellow dorsolateral stripes. Can alter intensity of colour and pattern, depending on temperature and reproductive status. Underside white with parallel grey stripes.

DISTRIBUTION Central Australia, between Windorah, QLD, and Alice Springs, NT, and south to top of the Eyre Peninsula, SA. **HABITAT AND HABITS** Found in sand-ridge deserts, semi-arid grassland and mallee shrubland. Diurnal. Often seen sitting on clumps of canegrass and spinifex. Sleeps on canegrass stems near the ground. Feeds on small invertebrates. Lays 1–6 eggs per clutch.

Long-nosed Dragon ■ *Gowidon longirostris* SVL 15cm

DESCRIPTION Body lightly built. Pale grey to brown, usually with series of paler broad blotches coalescing to form band along either side of spine from shoulder on to hips. Breeding males can be black with white stripe extending from snout along flanks. Can alter intensity of colour and pattern, depending on temperature and reproductive status. Short spines along vertebral ridge, forming low crest. **DISTRIBUTION** Arid Australia, from Geraldton to Broome, WA, east to Barkly Homestead through eastern NT, and south into

northern SA. **HABITAT AND HABITS** Found in dry open woodland, rocky gorges and tropical savannah. Diurnal. Often seen by day perched on trees and tops of boulders. Communicates with other dragons with head bobbing, press-ups and arm waves, hence the alternative name Ta-Ta Lizard. Feeds on invertebrates. At night may be seen sleeping on branches. Lays 4–12 eggs per clutch.

Water Dragon ■ *Intellagama lesueurii* SVL 25cm

DESCRIPTION Body medium in build. Two visually distinct subspecies. Female and juveniles any shade of grey or brown with flecks of black, white and yellow. Usually poorly defined darker bars across body. Tail barred with light bands. Breeding *I. l. howitti* males bluish-green with orange stripes on throat. *I. l. lesueurii* males white to grey with bright red chest and strong black bands. Both sexes have black temporal stripe edged with yellow. Can alter intensity of colour and pattern, depending on temperature and reproductive status. Spines along vertebral ridge, forming low crest. **DISTRIBUTION** *I. l. howitti* between Kangaroo Valley, NSW, through Canberra to Moe, VIC; *I. l. lesueurii* between Cooktown, down along east coast to Kangaroo Valley, NSW. Introduced populations around Melbourne, Werribee, VIC and Adelaide, SA. **HABITAT AND HABITS** Found in dry open woodland and forests along waterways. Often seen by day perched on trees and tops of boulders. Communicates with other dragons with head bobbing, press-ups and arm waves. Feeds on invertebrates. At night may be seen sleeping on branches. Lays 7–19 eggs per clutch.

I. l. leseuerii

I. l. howitti

Horner's Dragon ■ *Lophognathus horneri* SVL 10cm

DESCRIPTION Body medium in build. Pale grey to brown, usually with series of paler broad blotches coalescing to form band along either side of spine from shoulder on to hips. Breeding males can be black with white stripe extending from snout along flanks. White spot on tympanum. Can alter intensity of colour and pattern, depending on temperature and reproductive status. Short spines along vertebral ridge, forming low crest. **DISTRIBUTION** Tropical Australia, from Townsville, QLD, to Derby, WA. **HABITAT AND HABITS** Found in dry open woodland, rocky gorges and tropical savannah. Often seen by day perched on trees and fence posts. Communicates with other dragons with head bobbing, press-ups and arm waves, hence the alternative name Ta-Ta Lizard. Feeds on invertebrates. At night may be seen sleeping on branches. Lays 3–11 eggs per clutch.

Boyd's Forest Dragon ■ *Lophosaurus boydii* SVL 16cm

DESCRIPTION Body robust. Greenish-grey to brown with prominent sharp tubercles. Short spines along vertebral ridge. Blue to pink markings along flanks intersect with dark irregular blotches. Underside similar to dorsum. Some individuals have yellow flecks. **DISTRIBUTION** Wet Tropics region of north QLD, between Rossville and Paluma. **HABITAT AND HABITS** Restricted to rainforest. Shelters in leaf litter and beneath fallen epiphytic ferns. Diurnal. Often seen by day on trees, where it ambushes passing invertebrates. At night may be seen sleeping vertically on branches and small tree trunks. Lays 2–8 eggs per clutch.

Variations of species

Southern Angle-headed Dragon ■ *Lophosaurus spinipes* SVL 12cm

DESCRIPTION Body robust. Mottled greenish-grey to brown with prominent sharp tubercles. Short spines along vertebral ridge. Dark irregular flecks across body on

female. Underside similar to dorsum. Some individuals have yellow flecks. **DISTRIBUTION** Between Maleny, QLD, and Ourimbah, NSW. **HABITAT AND HABITS** Restricted to rainforest and closed forest. Shelters in leaf litter and beneath fallen epiphytic ferns. Diurnal. Often seen by day on small saplings about 80cm above the ground, where it ambushes passing invertebrates. Lays 2–8 eggs per clutch.

Thorny Devil ■ *Moloch horridus* SVL 12cm

DESCRIPTION Body robust. Covered in spines. Reddish, pale grey, yellow or brown, with series of broad blotches. Pair of pale yellow dorsolateral stripes. Underside white with series of grey and brown blotches that can be used to identify individuals. **DISTRIBUTION** Central Australia, between Windorah, QLD, to Alice Springs, NT, south to top of the

Eyre Peninsula, SA, and across to WA coastline between Geraldton and Pardoo. **HABITAT AND HABITS** Found in sand-ridge deserts, semi-arid grassland and mallee shrubland. Diurnal. Often seen moving with slow, jerky gait. Drinks water by brushing dew off plants, or standing in the rain or in a pool of water, using capillary attraction to wick water to mouth. Eats small black ants. Lays 1–7 eggs per clutch.

Eastern Bearded Dragon ■ *Pogona barbata* SVL 27cm
(Jew Lizard, Common Bearded Dragon)

DESCRIPTION Body robust. Pale grey to charcoal, yellow to any shade of brown. Underside white with grey occelli. Breeding males can have vivid black beard. Can alter intensity of colour and pattern, depending on temperature, stress and reproductive status. Upper body covered in tubercles. Tail has rings of enlarged tubercles and is often banded. Juveniles usually pale with complex pattern of grey to brown markings. **DISTRIBUTION** From Mareeba, QLD, through eastern Australia, into south-west VIC, and west across to

western edge of the Eyre Peninsula, SA. **HABITAT AND HABITS** Found in open woodland, heaths, grassland verges and forest verges. Diurnal. Often seen by day perched on trees or fence posts, or basking on bitumen roads. Communicates with other dragons with head bobbing and arm waves. Feeds on vegetation and invertebrates. Lays 3–32 eggs per clutch.

S Eipper

Downs Bearded Dragon ■ *Pogona henrylawsoni* SVL 15cm

DESCRIPTION Body robust. Pale grey to grey-brown. Underside white with grey occelli. Upper body covered in tubercles. Tail has rings of enlarged tubercles and is often banded. Juveniles usually pale with complex pattern of grey to brown markings. Can alter intensity

of colour and pattern, depending on temperature, stress and reproductive status. **DISTRIBUTION** From Croydon to Isisford, QLD. **HABITAT AND HABITS** Found in open woodland and Mitchell grassland on black soil. Diurnal. Hides under vegetation and in soil cracks. Feeds on vegetation and invertebrates. Lays 5–15 eggs per clutch.

R Francis

Western Bearded Dragon ■ *Pogona minor* SVL 27cm

DESCRIPTION Body robust. Pale grey to charcoal, yellow to any shade of brown.
Underside white with grey occelli. Upper body covered in tubercles. Tail often banded.
Juveniles usually pale with complex pattern of grey to brown markings. Beard poorly
developed. Can alter intensity of colour and pattern, depending on temperature, stress
and reproductive status. **DISTRIBUTION** From Alice Springs, NT, south to Port
Augusta, SA, and west across WA, excluding the Kimberley. Isolated population on the

Barkly Tableland. **HABITAT
AND HABITS** Found in open
woodland, heaths, grassland
verges, mallee, sand-ridge
deserts, mulga and claypans.
Diurnal. Often seen by day
perched on trees and fence
posts. Communicates with other
dragons with head bobbing and
arm waves. Feeds on vegetation
and invertebrates. Lays 3–32 eggs
per clutch.

Central Bearded Dragon ■ *Pogona vitticeps* SVL 24cm
(Inland Bearded Dragon)

DESCRIPTION Body robust. Pale grey to charcoal, yellow or orange, to any shade of
brown. Underside white with grey occelli. Breeding males can have vivid black beard.
Upper body covered in tubercles. Juveniles usually pale with complex pattern of grey to
brown markings. Can alter intensity of colour and pattern, depending on temperature,
stress and reproductive status. **DISTRIBUTION** Across the Eyre Basin from the Barkly

Tableland, NT, south to Eyre
Peninsula, SA, east into western
VIC, and north through NSW and
QLD. **HABITAT AND HABITS**
Found in open woodland, mallee,
sand-ridge deserts, mulga and
claypans. Diurnal. Often seen
by day perched on trees or fence
posts, or basking on bitumen
roads. Communicates with other
dragons with head bobbing and
arm waves. Feeds on vegetation
and invertebrates. Lays 3–43 eggs
per clutch.

Mountain Heath Dragon ■ *Rankinia diemensis* SVL 7cm

DESCRIPTION Body robust. Pale grey to reddish-brown, usually with series of paler broad blotches coalescing to form band along either side of spine from shoulder to hips. Series of rows of tubercles across upper body and legs. Mouth lining yellow and blue tongue. Can alter intensity of colour and pattern, depending on temperature, stress and reproductive status. Probably a species complex. **DISTRIBUTION** From upland areas of eastern NSW, to south-west VIC and TAS. **HABITAT AND HABITS** Found in open woodland, heaths and grassland verges. Diurnal. Often seen by day perched on low branches and small stones. Feeds on invertebrates. Lays 2–8 eggs per clutch.

S Eipper

Swampland Lashtail ■ *Tropicagama temporalis* SVL 13cm

DESCRIPTION Body medium in build. Pale grey to brown, usually with series of paler broad blotches coalescing to form band along either side of spine from shoulder on to hips. Hips, tail and back legs usually rusty-red to pale brown. Breeding males can be black with white stripe extending from snout along flanks. White spot on tympanum. Can alter intensity of colour and pattern, depending on temperature and reproductive status. Strong nuchal crest with deep, pointed scales. **DISTRIBUTION** Tropical Australia, from Townsville, QLD, to Derby, WA. **HABITAT AND HABITS** Found along water courses in tropical savannah and on to floodplains. Diurnal. Often seen by day perched on trees and fence posts. Communicates with other dragons with head bobbing, press-up and arm waves, hence the alternative name Ta-Ta Lizard. Feeds on invertebrates. At night may be seen sleeping on branches. Lays 3–11 eggs per clutch.

H Cogger

Condamine Earless Dragon ■ *Tympanocryptis condaminensis* SVL 6cm

DESCRIPTION Body robust. Pale grey to dark grey-brown with complex pattern of black, white and grey markings. Tail has series of pale bands. Often thin white stripes along body. Prominent tubercles across body and legs. Underside white. Males can develop orange

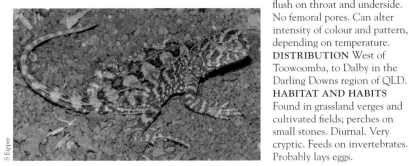

flush on throat and underside. No femoral pores. Can alter intensity of colour and pattern, depending on temperature. **DISTRIBUTION** West of Toowoomba, to Dalby in the Darling Downs region of QLD. **HABITAT AND HABITS** Found in grassland verges and cultivated fields; perches on small stones. Diurnal. Very cryptic. Feeds on invertebrates. Probably lays eggs.

Smooth-snouted Earless Dragon ■ *Tympanocryptis intima* SVL 6cm

DESCRIPTION Body rotund and depressed. Smooth scaled with sparsely distributed tubercles. Reddish-brown to grey, with complex pattern of black, white and grey markings. Tail usually has series of pale bands. Underside white in both sexes Can alter intensity of colour and pattern, depending on temperature. No femoral pores. **DISTRIBUTION** Eyre Basin, from Karumba, through western QLD and NSW, to Lake Everard, central eastern SA. **HABITAT AND HABITS** Found on heavy stony soils, claypans and gibber desert. Diurnal. Usually seen perched on small stones. Feeds on invertebrates. Lays 2–5 eggs per clutch.

Lined Earless Dragon ■ *Tympanocryptis lineata* SVL 7cm

DESCRIPTION Body robust. Pale grey to dark grey-brown with complex pattern of black, white and grey markings. Tail has series of pale bands. Often thin white stripes along body. Underside white in both sexes. Prominent tubercles across body and legs. Can alter intensity of colour and pattern, depending on temperature. Single femoral pore per side.

DISTRIBUTION Southern ACT and neighbouring NSW. **HABITAT AND HABITS** Found on the few remnants of native grassland. Diurnal. Hides in spider burrows and under small rocks. In significant decline and currently threatened with extinction due to habitat loss and feral predation. Feeds on invertebrates. Lays 2–6 eggs per clutch.

S Eipper

Eyrean Earless Dragon ■ *Tympanocryptis tetraporophora* SVL 6cm

DESCRIPTION Body robust. Pale grey to dark grey-brown with complex pattern of black, white and grey markings. Tail has series of pale bands. Often thin white stripes along body. Underside white. Prominent tubercles across body and legs. Can alter intensity of colour and pattern, depending on temperature. Two femoral pores. Probably a species complex. **DISTRIBUTION** Across Australia, west of the GDR, from Goondiwindi north-west

to Gulf of Carpentaria and eastern NT, south through eastern SA and western NSW. **HABITAT AND HABITS** Found in open woodland, Mitchell grassland, sand-ridge desert, mulga and claypans. Diurnal. Often seen by day perched on low branches and small stones. Sits outside during heat of the day, modifying posture to dissipate heat. Feeds on invertebrates. Lays 2–6 eggs per clutch.

S Eipper

MONITORS

Ridge-tailed Monitor ■ *Varanus acanthurus* TL 70cm
(Spiny-tailed Monitor, Ackie)

DESCRIPTION Body robust. Top of body reddish to very dark brown with white, yellow or cream ocelli and spots. Most individuals have longitudinal stripes on top of neck. Underside cream. Tail has spinous scales with dark bands and light yellow scales. Probably a species complex. **DISTRIBUTION** QLD, from Winton, across northern Australia,

including Top End of the NT, to Carnarvon, WA. Dwarf form on Barrow Island, WA. **HABITAT AND HABITS** Found in stony ranges, red-sand deserts, tropical woodland and rocky outcrops, in spinifex, under boulders or man-made debris, or in crevices. Diurnal. Terrestrial. Feeds on spiders, insects and small lizards. Lays 2–18 eggs per clutch.

Black-spotted Spiny-tailed Monitor ■ *Varanus baritji* TL 70cm
(White's Monitor, Lemon-throated Monitor)

DESCRIPTION Body robust. Top of body reddish-brown to grey with small dark spots, sometimes with white-edged occelli. Pale-edged dark canthal stripe continues as temporal stripe. Some individuals have dark stripe from jaw to neck. Neck, limbs and tail flecked with white and dark brown. Underside white to fawn. Yellow throat. Species complex. *V. b. insulanicus* sometimes placed as subspecies of *V. acanthurus*, but more closely related to

V. baritji. Additional blue-striped form at Cape Crawford, NT. **DISTRIBUTION** Northern NT, from Limmen NP, to Adelaide River Hills area. **HABITAT AND HABITS** Found in rocky outcrops, stony hills and tropical woodland escarpments. Diurnal. Terrestrial. Occurs under rocks and logs, and inside rock crevices. Feeds on spiders, insects and small lizards. Lays 3–12 eggs per clutch.

Short-tailed Pygmy Monitor ■ *Varanus brevicauda* TL 25cm

DESCRIPTION Body robust. Top of body light yellowish-brown to reddish-brown, covered in fine dark brown, grey and pale flecking. Most individuals have dark temporal stripe. Underside white. **DISTRIBUTION** Arid central Australia, from Ethabuka Station, QLD, west into southern half of the NT and far northern SA, across to WA coastline between Shark Bay and Willare. **HABITAT AND HABITS** Found in hummock grassland, red-sand deserts and stony areas in spinifex and under rocks. Diurnal. Terrestrial. Feeds on small invertebrates, and small lizards and their eggs. Lays 2–5 eggs per clutch.

Bush's Monitor ■ *Varanus bushi* TL 35cm
(Pilbara Pygmy Monitor)

DESCRIPTION Body lightly built. Top of body greyish-brown fading to light grey on sides. Reddish-brown spots and bars on sides, head and limbs. Most individuals have dark brown temporal stripe. Tail grey with dark brown banding at base turning into longitudinal stripes. **DISTRIBUTION** Only the Pilbara region of WA, between Peedamulla Homestead, Mt Whaleback and Roebourne. **HABITAT AND HABITS** Occurs in hummock grassland, open woodland and mulga. Found in gum, mulga or desert oak trees, in cracks and crevices, and under bark. Diurnal. Arboreal. Also utilizes ground-shelter sites. Eats small invertebrates and small lizards. Lays 3–4 eggs per clutch.

Stripe-tailed Monitor ■ *Varanus caudolineatus* TL 30cm
(Short-tailed Monitor, Pygmy Stripe-tailed Monitor)

DESCRIPTION Body lightly built. Top greyish-brown with dark brown flecks covering head and body. Tail-base has dark brown spots that merge into longitudinal stripes towards

tip. Underside creamy-white with occasional dark brown flecks. **DISTRIBUTION** Central WA, between Exmouth and Kalbarri, west to Kalgoorlie, and north to Balfour Downs Station. **HABITAT AND HABITS** Occurs in hummock grassland, open woodland, mulga and mallee. Found in mulga or desert oak trees in cracks and crevices, and under bark. Diurnal. Arboreal. Feeds on insects, spiders and small lizards. Lays 2–6 eggs per clutch.

Mangrove Monitor ■ *Varanus chlorostigma* TL 1.5m

DESCRIPTION Body robust. Top of body dark green, grey, black or blue with gold to cream spots. Underside similar to dorsum. Tail laterally compressed for swimming. **DISTRIBUTION** Coastal north QLD, from Cape Melville to Pormpuraaw, and the NT from Groote Eylandt to Darwin. Also PNG and Indonesia. **HABITAT AND HABITS** Found in coastal forests, mangroves and neighbouring rainforest. Diurnal. Arboreal to semi-aquatic, feeding on large invertebrates, crustaceans, fish, other reptiles, birds, eggs and carrion. One of two monitor species with salt-excretion glands. Lays 2–10 eggs per clutch.

Blue-tailed Monitor ■ *Varanus doreanus* TL 1.75m
(Kalabeck's Monitor)

DESCRIPTION Body robust. Greenish-blue to dark grey with pale yellow-green to white
flecks and occelli. Tail has
double keel with blackish-grey
bands, the last third being
pale blue. Underside white to
cream. Throat marked with
dark brown to black. Tongue
yellow. **DISTRIBUTION**
Iron Range NP, Nesbit River,
Newcastle Bay and Lockerbie
Scrub, QLD. Also PNG.
HABITAT AND HABITS
Lives in coastal mangroves,
vine thickets and rainforests.
Diurnal. Semi-arboreal. Feeds
on birds, small mammals and
invertebrates. Lays eggs.

A Zimny

Striped Desert Monitor ■ *Varanus eremius* TL 50cm
(Pygmy Desert Monitor)

DESCRIPTION Body lightly built. Top of body light to dark reddish-brown with dark
brown to black flecking and streaks. Pale yellow to white spots sparsely cover body. Tail
has longitudinal stripes that alternate in colour from cream to brown. Black stripe from
eye to snout; light-coloured temporal stripe and light-coloured stripe from jaw to base of
forelimbs. Underside white
or cream. Juveniles more
brightly coloured than
adults. **DISTRIBUTION**
WA, between Geraldton and
Eighty Mile Beach, through
arid central Australia, to
Ceduna, SA, and Tennant
Creek, NT, across to Boulia,
QLD. **HABITAT AND
HABITS** Found in sandy
deserts in and around
spinifex. Diurnal. Feeds on
small lizards and insects.
Lays 2–6 eggs per clutch.

R Valentic

Perentie ■ *Varanus giganteus* TL 2.5m

DESCRIPTION Australia's largest lizard. Body pale brown to dark brown or grey, with yellow, cream or white spots forming transverse bands. Throat reticulated with black or grey over white, and tail-tip white or yellow. Underside white or cream. Juveniles more brightly coloured than adults. **DISTRIBUTION** WA, between Rothsay and Port Hedland,

through arid central Australia to Farina, SA, and Tanam, NT, across to Windorah and Kynuna, QLD. **HABITAT AND HABITS** Found in deserts, rocky hills, black-soil grassland and coastal heaths. Diurnal. Terrestrial, sheltering in burrows and small caves. Feeds predominantly on lizards, snakes and carrion; also hunts birds and mammals. Lays 5–13 eggs per clutch.

S Eipper

Pygmy Mulga Monitor ■ *Varanus gilleni* TL 40cm
(Gillen's Monitor)

DESCRIPTION Body lightly built. Grey to brown with banded and spotted grey sides. Head and limbs have dark brown flecking and reticulations. Usually dark streak behind each eye. Banding on tail-base turns into longitudinal stripes midway along length.

DISTRIBUTION Arid areas of inland Australia, from Eighty Mile Beach, WA, through inland WA, across southern half of the NT, and most of arid SA to Port Augusta. **HABITAT AND HABITS** Found in hummock grassland, open woodland, mulga and mallee. Diurnal. Arboreal. Shelters under bark or inside trees crevices. Feeds on insects and small lizards. Lays 2–12 eggs per clutch.

A. Elliott

Kimberley Rock Monitor ▪ *Varanus glauerti* TL 80cm
(Glauert's Monitor)

DESCRIPTION Body lightly built. Top of body reddish to very dark brown with lighter coloured spots forming bands on to flanks. Tail in both populations dark with yellow to white bands. Throat and underside white. Species complex, with names V. *balagardi* and V. *gija* both proposed for the eastern form. **DISTRIBUTION** Three populations, one in western Arnhem Land and neighbouring areas of Kakadu NP, central population from Bullo River in the NT to Kununurra, WA, and another in the western Kimberleys of WA. **HABITAT AND HABITS** Found in rocky gorges, escarpments and open woodland. Diurnal. Arboreal or saxicoline. Feeds on insects, spiders and small lizards. Lays 3–12 eggs per clutch.

S Eipper

Twilight Monitor ▪ *Varanus glebopalma* TL 1.1m
(Black-palmed Monitor, Black-footed Monitor, Long-tailed Rock Monitor)

DESCRIPTION Body lightly built. Top of body dark grey to black with light brown flecking that forms reticulum on flanks. Flecking forms spots on limbs. Top of tail black; bottom and tip yellow to white. Lips barred. Throat white with light purplish-brown reticulum. Ventral surface white with light purplish-brown banding. Palms and soles dark. Scales on feet protrude, assisting with rock climbing. **DISTRIBUTION** Cloncurry, QLD, west across Top End of the NT and Kimberleys of WA. **HABITAT AND HABITS** Saxicoline monitor found in rocky gorges, escarpments and open woodland. Diurnal to crepuscular. Feeds on insects, birds, frogs, reptiles and small mammals. Lays 1–8 eggs per clutch.

R Francis

Sand Monitor ■ *Varanus gouldii* TL 1.6m
(Gould's Monitor, Racehorse Goanna, Bungarra)

DESCRIPTION Body robust. Ranges from yellow to black, with white, yellow or brown ocelli on dorsum. Dark streak behind eye outlined in yellow or cream. Throat has grey streaks. Tail banded, with posterior quarter white or yellow. Ventral surface cream to yellow with or without grey flecks. **DISTRIBUTION** Most parts of dry Australian mainland

except for south-east of the country. **HABITAT AND HABITS** Found in open woodland, heaths, mallee, brigalow, mulga, spinifex-covered sand plains and sandy deserts. Diurnal. Shelters in burrows and hollow logs, and occasionally climbs trees. Feeds on invertebrates, insects, eggs, birds, small mammals, reptiles and carrion. Lays 3–20 eggs per clutch.

Southern Pilbara Rock Monitor ■ *Varanus hamersleyensis* TL 50cm

DESCRIPTION Body lightly built. Reddish-brown above. Head and neck have darker flecking. Banding on neck in some individuals. Dark brown spots with pale centre on back. Limbs have spots. Tail has narrow grey and dark reddish-brown banding. Dark temporal streak in some individuals. Underside pale with grey flecking or banding. **DISTRIBUTION** Only the Pilbara region, WA, between Mt Rica and Newman. **HABITAT AND HABITS** Found in rocky outcrops and cliff faces. Diurnal. Saxicoline. Occurs in crevices and cavities or under boulders, where it feeds on invertebrates and small lizards. Lays 3–6 eggs per clutch.

Canopy Monitor ■ *Varanus keithhornei* TL 75cm

DESCRIPTION Body lightly built. Top of body dull bluish-grey to black, often with
yellow to cream scales
that form poorly defined
cross-bands. Prehensile
tail used during climbing
through the canopy.
DISTRIBUTION
McIlwraith Range and
Iron Range of Cape
York Peninsula, QLD.
HABITAT AND HABITS
Occurs in mangroves,
rainforest, closed forest
and vine thickets.
Diurnal. Arboreal. Found
in tree hollows or while
basking on trees. Feeds on
insects. Lays 2–4 eggs per
clutch.

Long-tailed Rock Monitor ■ *Varanus kingorum* TL 40cm
(Pygmy Rock Monitor, Kings Monitor, Kings Rock Monitor)

DESCRIPTION Body lightly built. Top of body reddish-brown with fine black to grey
spots. In some individuals spots align to form reticulate pattern. Underside white to
yellow with dark brown
spots. **DISTRIBUTION**
Warmun in the eastern
Kimberley, WA, and to
Timber Creek in north-
west NT. **HABITAT
AND HABITS** Occurs
in rocky gorges and
limestone outcrops with
open woodland. Diurnal.
Saxicoline. Found in
rock crevices, under rocks
and in spinifex. Feeds on
insects and small lizards.
Lays 2–4 eggs per clutch.

Merten's Water Monitor ■ *Varanus mertensi* TL 1.5m
(Water Goanna)

DESCRIPTION Body robust. Top of body greenish-brown to black with many small, cream to yellow spots. Sides of head often flushed with yellow. Lower lip has grey barring. Some individuals have lower bluish powdering with orange streaks. Underside cream to yellow with some dark flecking, sometimes forming incomplete cross-bands. **DISTRIBUTION** Kimbolton in the Kimberleys of WA, through Top End of the NT, across to Cape York,

QLD. Isolated population near Clermont, QLD. **HABITAT AND HABITS** Found in swamps, lagoons and rivers, often in rocky gorges that cut through open woodland. Diurnal. Semi-aquatic. Often seen basking on rocks or in trees overhanging water. Feeds mainly on crustaceans, but also eats frogs, fish, lizards, small mammals, snakes, birds, eggs and carrion. Lays 3–14 eggs per clutch.

Mitchell's Water Monitor *Varanus mitchelli* TL 90cm
(Mitchell's Goanna)

DESCRIPTION Body lightly built. Top of body orange, brown to bluish-black, with many cream or yellow spots with black centres. Yellow flecking on head; yellow stripe behind eyes. Throat yellow with black flecking. Black barring on lower lip. Limbs and tail black with yellow flecking. Underside yellow with dark spotting. **DISTRIBUTION** From Derby, WA, through the Kimberley, across Top End of the NT, to the QLD border. Single record from Gregory River, near

Riversley, QLD. **HABITAT AND HABITS** Usually found near freshwater swamps, lagoons and rivers. Diurnal. Semi-aquatic. Also uses rocky areas on coastlines, basking on rocks. Shelters in rock crevices and tree hollows. Feeds on spiders, insects, crustaceans, frogs, fish, lizards and small rodents. Lays 3–27 eggs per clutch.

Yellow-spotted Monitor *Varanus panoptes* TL 1.6m

DESCRIPTION Body robust. Pale brown to dark brown or grey, with or without yellow, cream or white spots with wide black bands; smaller yellow spots often arranged in transverse rows. Throat spotted or flecked. Tail yellow and usually banded with brown to tip. **DISTRIBUTION** V. *p. panoptes* from QLD to northern WA, and northern NT; V. *p. rubidus* from the Pilbara to the Goldfields in WA; V. *p. horni* in PNG. **HABITAT AND HABITS** Found across all habitats in its range, with the exception of rainforest. Diurnal. Terrestrial. Shelters in burrows and small caves. Feeds on lizards, snakes and carrion, and sea-turtle eggs; also hunts birds and mammals. Lays 6–19 eggs per clutch.

S Eipper

Northern Pilbara Rock Monitor *Varanus pilbarensis* TL 50cm

DESCRIPTION Body lightly built. Reddish-brown with irregular pale yellow-grey spots aligned to form cross-bands. Head and neck have darker flecking. Some individuals have banding on neck. Dark brown spots with pale centres on back. Limbs have spots. Tail blackish with yellow to white bands. Dark temporal streak in some individuals. Underside pale with grey flecking or banding. **DISTRIBUTION** Only the Pilbara region, WA, from Karratha to Goldsworthy, and south to Knox Gorge. **HABITAT AND HABITS** Occurs in rocky outcrops and on cliff faces. Diurnal. Saxicoline. Found in crevices or cavities, or under boulders, where it feeds on invertebrates and small lizards. Lays 2–5 eggs per clutch.

M Summerville

Emerald Monitor *Varanus prasinus* TL 95cm
(Wynsis, Green Tree Monitor, Green Lace Lizard)

DESCRIPTION Body lightly built. Light to emerald-green above with black flecking that forms cross-bands. Narrow longitudinal lines on throat. Underside white, yellow or pale green. Soles of feet have enlarged scales, and tail prehensile. **DISTRIBUTION** Moa

and Murray Islands in the Torres Strait, QLD. Also PNG and Indonesia; unconfirmed reports on Cape York Peninsula. **HABITAT AND HABITS** Found in rainforests, palm forests, mangroves and lagoons. Diurnal. Arboreal. Usually seen in trees and on vines. Feeds on invertebrates, and small lizards, mammals and birds. Lays 2–8 eggs per clutch.

Blunt-spined Monitor *Varanus primordius* TL 30cm
(Northern Blunt-nosed Monitor, Northern Ridge-tailed Monitor)

DESCRIPTION Body robust. Usually light reddish-brown to dark greyish-brown with dark brown flecking. Dark temporal stripe. Underneath white to cream with or without dark flecks. Tail has blunt, spiny projections. **DISTRIBUTION** Top End region of the NT,

from between Chambers River, to Daly River and Hotham. **HABITAT AND HABITS** Occurs in rocky outcrops with tropical open woodland. Diurnal but can be crepuscular in warm weather. Found in rock crevices and soil cracks, and beneath rocks, fallen timber and man-made debris. Feeds on insects, spiders, centipedes, small lizards and lizard eggs. Lays 1–6 eggs per clutch.

Heath Monitor *Varanus rosenbergi* TL 1.5m
(Rosenberg's Monitor)

DESCRIPTION Body robust. Blackish above with fine dots of white or yellow. About 15 narrow black bands from neck down body. Black temporal stripe with white edging. Tail has blackish-brown and light yellow banding. Lips barred. Probably a species complex. **DISTRIBUTION** Three separate populations. Western population in southern WA, from Perth to Eucla; central population from southern Eyre Peninsula, through coastal SA including Kangaroo Island, and into western VIC to Mt Arapiles; eastern population from Tallangatta, VIC, across the GDR, into the Sydney Basin, extending north to Yengo NP. **HABITAT AND HABITS** Occurs in coastal heaths, heathland, sclerophyll forests and humid wetlands. Diurnal. Terrestrial. Found in rock crevices, hollow logs and burrows. Feeds on insects, frogs, reptiles, birds, small mammals and carrion. Lays 3–19 eggs per clutch.

Spotted Tree Monitor *Varanus scalaris* TL 65cm
(Banded Tree Monitor)

DESCRIPTION Body lightly built. Very variable in appearance. Most individuals brown, grey to blackish above, with white to yellow spotting and flecking. Black temporal stripe edged in white. Underside white to light grey or yellow, with dark flecking. Limbs spotted with white. Tail grey to black, with or without light banding. Species complex; *V. similis* sometimes used for Top End animals, *kuranda* for Wet Tropics animals, and *pelewensis* for western Cape York and Gulf animals. **DISTRIBUTION** Beagle Bay, WA, across northern Australia, to Airlie Beach, QLD. Possibly isolated population in the Cooper Creek Drainage. Also PNG. **HABITAT AND HABITS** Found in savannah woodland to rainforests. Diurnal. Arboreal. Seen basking on tree trunks and branches. Shelters in hollow logs, under tree bark and occasionally in rock crevices. Feeds on insects, small lizards, frogs, birds, eggs and fledglings. Lays 3–15 eggs per clutch.

Kuranda *form*

Similis *form*

Rusty Monitor *Varanus semiremex* TL 60cm

DESCRIPTION Body robust. Head and forebody usually reddish orange-brown or greyish-brown, with dark spotting and flecking that forms thin reticulum. Juveniles dark with

light-coloured spots. Underside yellow to grey with darker markings. **DISTRIBUTION** Coastal QLD, from Gladstone to Weipa. Some populations extend inland along river systems. **HABITAT AND HABITS** Found in coastal mangroves, swamps and surrounding freshwater river systems. Diurnal. Arboreal. Shelters in tree hollows and rotten fence posts, and among tree roots. Feeds mainly on fish, frogs, crabs, insects and small lizards. Lays 2–21 eggs per clutch.

Dampier Peninsula Monitor *Varanus sparnus* TL 26cm

DESCRIPTION Body robust. Top of body light yellowish-brown to pinkish-brown. Head, limbs and body have dark brown and light yellow flecking. Most individuals have dark temporal stripe. Underside white to cream. **DISTRIBUTION** Restricted to the Dampier Peninsula of southern Kimberley, WA. **HABITAT AND HABITS** Lives in hummock grassland and red sandy deserts. Diurnal. Terrestrial. Shelters in spinifex and under rocks. Feeds on small invertebrates, and small lizards and their eggs.

Spencer's Monitor *Varanus spenceri* TL 1.35m

DESCRIPTION Body robust. Very variable in colour, ranging from pale cream, grey to reddish-brown. Body has dark bands, flecks or cream spots. Lips barred, and tail has dark bands. Underside cream to grey with dark flecks. **DISTRIBUTION** QLD, from Jundah up to Nelia, and across to north-west of Tennant Creek, NT. **HABITAT AND HABITS** Lives in black-soil and arid plains, and grassland. Diurnal. Terrestrial. Found in cracks in clay and cavities under rocks. Feeds on insects, mammals, reptiles and carrion. Can go through significant weight loss, becoming emaciated during periods of drought, before recovering rapidly post floods. Lays 4–35 eggs per clutch.

Storr's Monitor *Varanus storri* TL 40cm

DESCRIPTION Body robust. Top of body red to brown. Most individuals have ocelli or flecking. Throat creamish to yellow. Usually a dark temporal stripe. Tail short with long, spiny projections. Two subspecies. *V. s. storri,* in western half of range, lacks enlarged scales on undersides of legs; *V. s. ocreatus,* in eastern half of range, has enlarged scales on undersides of legs. **DISTRIBUTION** From Charters Towers, QLD, across northern Australia, to the Kimberley, WA. **HABITAT AND HABITS** Found in rocky outcrops, rocky grassland and open woodland. Diurnal. Terrestrial. Often found under rocks, fallen timber and man-made debris; occasionally in burrows dug beneath grass tussocks. Feeds on insects and small lizards. Lays 1–6 eggs per clutch.

Black-headed Monitor *Varanus tristis* TL 80cm
(Black-tailed Monitor, Freckled Monitor, Mournful Goanna)

DESCRIPTION Body lightly built. Very variable in colour. Pale grey, yellow to dark brown, or black above. Body usually scattered with white to reddish ocelli, which fade with age. Some individuals have black heads. Legs and tail black, with or without irregular cream spots. Sometimes irregular bands on tail. Underside pale yellow to white. **DISTRIBUTION** Across Australia north of Perth, WA, to Coonamble, NSW; west of the GDR in south

of range, crossing it at Kingaroy, QLD. **HABITAT AND HABITS** Found in sandy deserts, grassland, brigalow and woodland. Diurnal. Arboreal. Seen basking on tree trunks, branches and rock faces. Shelters in hollow logs, under tree bark and in rock crevices; in some parts of range in roofs of houses. Feeds mainly on small lizards, and occasionally frogs, insects and bird fledglings. Lays 3–15 eggs per clutch.

Lace Monitor *Varanus varius* TL 2.4m

DESCRIPTION Australia's heaviest lizard. Body robust. Two distinct colour forms. Black to dark grey above with yellow to white spots often forming bands. Tail banded. Head and neck can have blue flecking. Other colour form known as the Bell's phase, black with broad yellow or white bands. Tail-tip in both forms usually white or yellow. **DISTRIBUTION** Eastern Australia, from Port Augusta, SA, across VIC and NSW, and north to Cooktown, QLD. **HABITAT AND HABITS** Found in many ecosystems, from rainforest to mallee, as well as open woodland, brigalow, grassland and coastal heaths. Diurnal. Arboreal. Shelters in burrows, hollowed out trees, termite mounds and caves. Adults predominantly feed on mammals, birds and carrion; occasionally reptiles, fish and insects. Also readily takes food from people at camp grounds and picnic areas. Lays 3–15 eggs per clutch.

Normal form

Bell's phase

CHECKLIST OF THE LIZARDS OF AUSTRALIA

Taxonomy follows Cogger 2018, with the exception of the addition of newly described taxa. For each species, an 'x' indicates presence in a particular state or territory. An '*' refers to species described in the book.

State or territory abbreviations:

NSW	New South Wales (including ACT)
NT	Northern Territory
Q	Queensland
S	South Australia
T	Tasmania
V	Victoria
W	Western Australia

Abbreviations of IUCN Red List status:

EX	Extinct
CR	Critically Endangered
EN	Endangered
VU	Vulnerable
NT	Near Threatened

Common English Name	Scientific Name	Qld	NSW	Vic	Tas	SA	WA	NT	Islands	IUCN
Geckos (Carphodactylidae)										
Chameleon Gecko*	Carphodactylus laevis	x								LC
Centralian Knob-tailed Gecko*	Nephrurus amyae						x	x		LC
Prickly Knob-tailed Gecko*	Nephrurus asper	x								LC
Northern Banded Knob-tailed Gecko	Nephrurus cinctus						x			
Pernatty Knob-tailed Gecko	Nephrurus deleani					x				LC
Pale Knob-tailed Gecko*	Nephrurus laevissimus	x				x	x	x		LC
Smooth Knob-tailed Gecko*	Nephrurus levis levis	x	x			x	x	x		LC
Northern Knob-tailed Gecko	Nephrurus sheai						x	x		LC
Starred Knob-tailed Gecko*	Nephrurus stellatus					x	x			LC
Mid-line Knob-tailed Gecko	Nephrurus vertebralis						x			LC
Southern Banded Knob-tailed Gecko*	Nephrurus wheeleri						x			LC
McIlwraith Leaf-tailed Gecko	Orraya occulta	x								VU
Mt Elliot Leaf-tailed Gecko	Phyllurus amnicola	x								NT
Ringed Thin-tailed Gecko	Phyllurus caudiannulatus	x								NT
Chapion's Leaf-tailed Gecko	Phyllurus championae	x								NT
Gulbaru Leaf-tailed Gecko	Phyllurus gulbaru	x								EN
Mt Blackwood Broad-tailed Gecko*	Phyllurus isis	x								VU
Oakview Leaf-tailed Gecko*	Phyllurus kabikabi	x								CR
Eungella Broad-tailed Gecko*	Phyllurus nepthys	x								LC
Mount Ossa Leaf-tailed Gecko	Phyllurus ossa ossa	x								LC
Pinnacles Leaf-tailed Gecko	Phyllurus pinnaclensis	x								NA
Broad-tailed Gecko*	Phyllurus platurus		x							LC
Northern Leaf-tailed Gecko*	Saltuarius cornutus	x								LC
Cape Melville Leaf-tailed Gecko	Saltuarius eximius	x								EN
Kate's Leaf-tailed Gecko	Saltuarius kateae		x							LC
Moritz's Leaf-tailed Gecko*	Saltuarius moritzi		x							LC
Rough-throated Leaf-tailed Gecko*	Saltuarius salebrosus	x								LC
Southern Leaf-tailed Gecko	Saltuarius swaini	x	x							LC
Granite Belt Leaf-tailed Gecko	Saltuarius wyberba	x	x							LC
Common Thick-tailed Gecko*	Underwoodisaurus milii	x	x	x		x	x	x		LC

Common English Name	Scientific Name	Qld	NSW	Vic	Tas	SA	WA	NT	Islands	IUCN
Pilbara Barking Gecko	Underwoodisaurus seorsus						x			LC
Granite Belt Thick-tailed Gecko*	Uvidicolus sphyrurus	x	x							LC
Geckos (Diplodactylidae)										
Clouded Gecko*	Amalosia jacovae	x	x							LC
Lesuer's Velvet Gecko*	Amalosia lesueurii	x	x							LC
Slim Velvet Gecko	Amalosia obscura						x			LC
Zigzag Velvet Gecko*	Amalosia rhombifer	x	x				x	x		LC
Central Uplands Clawless Gecko*	Crenadactylus horni					x		x		LC
Northern Clawless Gecko	Crenadactylus naso	x					x	x		LC
Western Clawless Gecko	Crenadactylus occidentalis						x			LC
South-western Clawless Gecko	Crenadactylus ocellatus						x			LC
Pilbara Clawless Gecko	Crenadactylus pilbarensis						x			LC
South-west Kimberley Clawless Gecko	Crenadactylus rostralis						x			LC
Cape Range Clawless Gecko	Crenadactylus tuberculatus						x			LC
Eastern Deserts Fat-tailed Gecko*	Diplodactylus ameyi	x	x							LC
Gulf Fat-tailed Gecko	Diplodactylus barraganae	x				x				LC
Western Fat-tailed Gecko	Diplodactylus bilybara						x			LC
South Coast Gecko	Diplodactylus calcicolus					x	x			LC
Cape Range Gecko	Diplodactylus capensis						x			NT
Variable Fat-tailed Gecko	Diplodactylus conspicillatus	x				x	x	x		LC
Kimberley Fat-tailed Gecko	Diplodactylus custos					x	x			LC
Lake Disappointment Ground Gecko	Diplodactylus fulleri						x			VU
Ranges Stone Gecko	Diplodactylus furcosus		x	x		x				LC
Northern Pilbara Beak-faced Gecko	Diplodactylus galaxias						x			LC
Mesa Gecko*	Diplodactylus galeatus					x		x		LC
Western Stone Gecko	Diplodactylus granariensis granariensis						x			
Northern Fat-tailed Gecko	Diplodactylus hillii							x		LC
Kenneally's Gecko	Diplodactylus kenneallyi						x			DD
Kluge's Gecko*	Diplodactylus klugei						x			LC
Desert Fat-tailed Gecko	Diplodactylus laevis	x				x	x	x		LC
Speckled Stone Gecko	Diplodactylus lateroides						x			LC
Pilbara Stone Gecko	Diplodactylus mitchelli						x			LC
Cloudy Stone Gecko	Diplodactylus nebulosus						x			LC
Ornate Gecko	Diplodactylus ornatus						x			LC
Eastern Fat-tailed Gecko*	Diplodactylus platyurus	x								LC
Spotted Sandplain Gecko	Diplodactylus polyophthalmus						x			VU
Pretty Gecko*	Diplodactylus pulcher					x	x			LC
Southern Pilbara Beak-faced Gecko	Diplodactylus savagei						x			LC
Tessellated Gecko*	Diplodactylus tessellatus	x	x	x		x		x		LC
Eastern Stone Gecko*	Diplodactylus vittatus	x	x	x		x				LC
Desert Wood Gecko	Diplodactylus wiru					x	x			LC
Reticulated Velvet Gecko*	Hesperoedura reticulata						x			LC
White-spotted Ground Gecko	Lucasium alboguttatum						x			LC
Southern Sandplain Gecko	Lucasium bungabinna					x	x			LC
Gibber Gecko*	Lucasium byrnei	x	x			x		x		LC
Beaded Gecko*	Lucasium damaeum	x	x	x		x	x	x		LC
Pale-striped Ground Gecko	Lucasium immaculatum	x				x				LC
Gilbert Ground Gecko	Lucasium iris	x								NA
Main's Ground Gecko	Lucasium maini						x			LC
Yellow-snouted Ground Gecko*	Lucasium occultum							x		EN
Mottled Ground Gecko	Lucasium squarrosum						x			LC
Box-patterned Gecko*	Lucasium steindachneri	x	x			x				LC
Crowned Gecko*	Lucasium stenodactylum	x	x			x	x	x		LC
Pilbara Ground Gecko	Lucasium wombeyi						x			LC

Common English Name	Scientific Name	Qld	NSW	Vic	Tas	SA	WA	NT	Islands	IUCN
Robust Velvet Gecko*	Nebulifera robusta	x	x							LC
Silver-eyed Velvet Gecko	Oedura argentea	x								NA
Beautiful Velvet Gecko*	Oedura bella	x						x		LC
Northern Velvet Gecko	Oedura castelnaui	x								LC
Inland Marbled Velvet Gecko*	Oedura cincta	x	x			x		x		LC
Northern Spotted Velvet Gecko*	Oedura coggeri	x								LC
Elegant Velvet Gecko*	Oedura elegans	x	x							NA
Fringe-toed Velvet Gecko*	Oedura filicipoda						x			LC
Western Marbled Velvet Gecko	Oedura fimbria						x			LC
Jewelled Velvet Gecko	Oedura gemmata							x		LC
Gracile Velvet Gecko	Oedura gracilis					x	x			LC
Quinkan Gecko	Oedura jowalbinna	x								NT
Arcadia Velvet Gecko	Oedura lineata	x								NA
Mereenie Velvet Gecko	Oedura luritja							x		LC
Northern Marbled Velvet Gecko*	Oedura marmorata							x		LC
Ocellated Velvet Gecko	Oedura monilis	x								LC
Limestone Range Velvet Gecko	Oedura murrumanu						x			LC
Groote Eylandt Marbled Velvet Gecko	Oedura nesos							x		NA
Ornate Velvet Gecko	Oedura picta	x	x							NA
Southern Spotted Velvet Gecko*	Oedura tryoni	x	x							LC
Giant Tree Gecko*	Pseudothecadactylus australis	x								LC
Western Giant Cave Gecko	Pseudothecadactylus cavaticus						x			LC
Northern Giant Cave Gecko*	Pseudothecadactylus lindneri							x		LC
Border Beaked Gecko	Rhynchoedura angusta	x	x			x				LC
Eyre Basin Beaked Gecko	Rhynchoedura eyrensis	x	x			x		x		LC
Brigalow Beaked Gecko*	Rhynchoedura mentalis	x								LC
Eastern Beaked Gecko*	Rhynchoedura ormsbyi	x	x	x						LC
Western Beaked Gecko	Rhynchoedura ornata					x	x	x		LC
Northern Beaked Gecko	Rhynchoedura sexapora						x	x		LC
Thorn-tailed Gecko	Strophurus assimilis					x	x			LC
Congoo Gecko	Strophurus congoo	x								DD
Jewelled Gecko*	Strophurus elderi	x	x			x	x	x		LC
Arnhem Phasmid Gecko	Strophurus horneri							x		VU
Southern Spiny-tailed Gecko*	Strophurus intermedius		x	x		x	x	x		LC
Southern Phasmid Gecko	Strophurus jeanae						x	x		LC
Kristin's Spiny-tailed Gecko*	Strophurus krisalys	x								LC
McMillan's Striped Gecko	Strophurus mcmillani						x			LC
Robust Striped Gecko	Strophurus michaelseni						x			LC
Exmouth Spiny-tailed Gecko	Strophurus rankini						x			LC
Robinson's Striped Gecko	Strophurus robinsoni						x	x		LC
Soft Spiny-tailed Gecko	Strophurus spinigerus spinigerus						x			LC
Western Spiny-tailed Gecko	Strophurus strophurus						x			LC
Northern Phasmid Gecko*	Strophurus taeniatus	x					x	x		LC
Golden-tailed Gecko*	Strophurus taenicauda taenicauda	x								LC
Golden-eyed Gecko	Strophurus trux	x								VU
Western-shield Spiny-tailed Gecko*	Strophurus wellingtonae						x			LC
Eastern Spiny-tailed Gecko*	Strophurus williamsi	x	x	x		x				LC
Mount Augustus Striped Gecko	Strophurus wilsoni						x			LC
Geckos (Gekkonidae)										
Nullarbor Marbled Gecko	Christinus alexanderi					x	x			LC
Lord Howe Island Gecko	Christinus guentheri								x	VU
Marbled Gecko*	Christinus marmoratus		x	x		x	x			LC
Pascoe River Ring-tailed Gecko	Cyrtodactylus adorus	x								LC
Iron Range Ring-tailed Gecko	Cyrtodactylus hoskini	x					x			LC
Kimberley Bent-toed Gecko	Cyrtodactylus kimberleyensis						x			DD

Common English Name	Scientific Name	Qld	NSW	Vic	Tas	SA	WA	NT	Islands	IUCN
Southern Ring-tailed Gecko*	Cyrtodactylus mcdonaldi	x								LC
McIlwraith Ring-tailed Gecko	Cyrtodactylus pronarus	x								LC
Christmas Island Forest Gecko	Cyrtodactylus sadleiri								x	EN
Cooktown Ring-tailed Gecko*	Cyrtodactylus tuberculatus	x								LC
East Arnhem Land Gehyra	Gehyra arnhemica							x		NA
Northern Dtella*	Gehyra australis							x		LC
Short-tailed Dtella	Gehyra baliola	x								LC
Borroloola Dtella	Gehyra borroloola	x						x		LC
Relictual Karst Gehyra	Gehyra calcitectus						x			NA
North-west Cape Gehyra	Gehyra capensis						x			LC
Chain-backed Dtella*	Gehyra catenata	x								LC
West Kimberley tree Gehyra	Gehyra chimera						x			NA
Western Cryptic Gehyra	Gehyra crypta						x			LC
Dubious Dtella*	Gehyra dubia	x	x					x		LC
Einasleigh Rock Dtella	Gehyra einasleighensis	x								LC
Amber Rock Dtella	Gehyra electrum	x								NA
Hamersley Range Spotted Gehyra	Gehyra fenestrula						x			LC
Small-spotted Mid-West Rock Gehyra	Gehyra finipunctata						x			LC
Plain Tree Gehyra	Gehyra gemina						x	x		NA
Kimberley Karst Gecko	Gehyra girloorloo						x			LC
Kimberley Granular-toed Gecko	Gehyra granulum						x			LC
Northern Pilbara Cryptic Gehyra	Gehyra incognita						x			LC
Bungle Bungle Dtella	Gehyra ipsa						x			NA
Robust Termitaria Gecko	Gehyra kimberleyi						x	x		LC
King's Dtella	Gehyra koira						x	x		LC
Litchfield Rock Gehyra	Gehyra lapistola							x		NA
Gulf Tree Gehyra	Gehyra lauta	x						x		NA
Southern Rock Dtella	Gehyra lazelli		x			x				LC
Large Pilbara Rock Gehyra	Gehyra macra						x			LC
Medium Pilbara Spotted Rock Gehyra	Gehyra media						x			LC
Small Pilbara Spotted Rock Gehyra	Gehyra micra						x			LC
Dwarf Dtella	Gehyra minuta							x		LC
Central Rock Dtella	Gehyra montium					x	x	x		LC
Moritz's Dtella	Gehyra moritzi	x						x		LC
Multi-pored Gecko	Gehyra multiporosa						x			LC
Stump-toed Gecko	Gehyra mutilata								x	LC
Northern Spotted Dtella*	Gehyra nana	x				x		x		LC
Kimberley Plateau Dtella	Gehyra occidentalis						x			LC
Pilbara Island Gehyra	Gehyra ocellata						x			DD
Arnhem Land Dtella	Gehyra pamela							x		LC
Litchfield Spotted Gecko	Gehyra paranana							x		LC
Burrup Peninsula Rock Gehyra	Gehyra peninsularis						x			DD
Pilbara Dtella	Gehyra pilbara						x	x		LC
Northern Kimberley Gecko	Gehyra pluraporosa						x			DD
Large-spotted Mid-West Rock Gehyra	Gehyra polka						x			LC
Southern Kimberley Spotted Gecko	Gehyra pseudopunctata						x			LC
Central Spotted Rock Dtella	Gehyra pulingka					x	x	x		LC
Spotted Rock Dtella	Gehyra punctata						x			LC
Purple Dtella	Gehyra purpurascens	x				x	x	x		LC
Robust Dtella*	Gehyra robusta	x						x		LC
Small Wedge-toed Gecko	Gehyra spheniscus						x			LC
Cresent-marked Pilbara Gehyra	Gehyra unguiculata						x			DD
Variegated Dtella*	Gehyra variegata					x	x			LC
Variable Dtella*	Gehyra versicolor	x	x	x		x		x		LC

Common English Name	Scientific Name	Qld	NSW	Vic	Tas	SA	WA	NT	Islands	IUCN
Crocodile-faced Dtella	Gehyra xenopus						x			LC
Asian House Gecko*	Hemidactylus frenatus	x	x				x	x		NA
Fox Gecko*	Hemidactylus garnotii		x							NA
Sri Lankan Spotted House Gecko	Hemidactylus parvimaculatus								x	NA
Flat-tailed House Gecko	Hemidactylus platyurus								x	NA
Black Pilbara Gecko	Heteronotia atra						x			LC
Bynoe's Gecko*	Heteronotia binoei	x	x	x		x	x	x		LC
Pale-headed Gecko*	Heteronotia fasciolatus	x						x		LC
North-west Prickly Gecko	Heteronotia planiceps						x	x		LC
Desert Cave Gecko	Heteronotia spelea						x			LC
Lister's Gecko	Lepidodactylus listeri								x	EX
Mourning Gecko*	Lepidodactylus lugubris	x						x		LC
Slender Mourning Gecko	Lepidodactylus pumilus	x								LC
Southern Cape York Nactus*	Nactus cheverti	x								LC
Northern Cape York Nactus	Nactus eboracensis	x								LC
Black Mountain Gecko*	Nactus galgajuga	x								LC
Pelagic Gecko	Nactus 'pelagicus'	x								LC
Flap-footed Lizards (Pygopodidae)										
Eared Worm-lizard	Aprasia aurita			x		x				NT
Batavia Coast Worm-lizard	Aprasia clairae						x			DD
Shark Bay Worm-lizard	Aprasia haroldi						x			LC
Red-tailed Worm-lizard*	Aprasia inaurita		x	x		x	x			LC
Gnaraloo Worm-lizard	Aprasia litorea						x			EN
Pink-tailed Worm-lizard*	Aprasia parapulchella		x	x						LC
Black-headed Worm-lizard	Aprasia picturata						x			DD
Flinder's Worm-lizard	Aprasia pseudopulchella					x				LC
Granite Worm-lizard	Aprasia pulchella						x			LC
Sand-plain Worm-lizard	Aprasia repens						x			LC
Ningaloo Worm-lizard	Aprasia rostrata						x			VU
Black-tipped Worm-lizard	Aprasia smithi						x			LC
Lined Worm-lizard*	Aprasia striolata			x		x	x			LC
Wicherina Worm-lizard	Aprasia wicherina						x			DD
Marble-faced Delma*	Delma australis		x	x		x	x	x		LC
Northern Delma	Delma borea	x					x	x		LC
Spinifex Delma*	Delma butleri	x	x	x		x	x	x		LC
Javelin Lizard?	Delma concinna concinna						x			LC
Desert Delma	Delma desmosa					x	x	x		LC
Pilbara Delma	Delma elegans						x			LC
Fraser's Delma*	Delma fraseri						x			LC
Side-barred Delma	Delma grayii						x			LC
Neck-barred Delma	Delma haroldi						x	x		LC
Heath Delm	Delma hebesa						x			LC
Striped Delma*	Delma impar		x	x						EN
Olive Delma*	Delma inornata	x	x	x		x				LC
Stripe-tailed Delma	Delma labialis	x								LC
Atherton Delma*	Delma mitella	x								LC
Adelaide Delma	Delma molleri					x				LC
Sharp-snouted Delma*	Delma nasuta	x				x	x	x		LC
Peace Delma	Delma pax						x			LC
Painted Delma	Delma petersoni					x	x			LC
Leaden Delma*	Delma plebeia	x	x							LC
North West Cape Delma	Delma tealei						x			EN
Black-necked Delma*	Delma tincta	x	x			x	x	x		LC
Collared Delma*	Delma torquata	x								LC
Burton's Snake-lizard*	Lialis burtonis	x	x	x		x	x	x		LC
Bronzeback*	Ophidiocephalus taeniatus					x		x		LC
Brigalow Scaly-foot*	Paradelma orientalis	x								LC
Edel Keeled Legless Lizard	Pletholax edelensis						x			NA

Common English Name	Scientific Name	Qld	NSW	Vic	Tas	SA	WA	NT	Islands	IUCN
Keeled Legless Lizard*	*Pletholax gracilis*						x			LC
Common Scaly-foot*	*Pygopus lepidopodus*	x	x	x		x	x			LC
Western Hooded Scaly-foot*	*Pygopus nigriceps*	x				x	x	x		LC
Robert's Scaly-foot	*Pygopus robertsi*	x								LC
Eastern Hooded Scaly-foot*	*Pygopus schraderi*	x	x	x		x		x		LC
Northern Hooded Scaly-foot	*Pygopus steelescotti*	x					x	x		LC
Skinks (Scincidae)										
Eastern Three-lined Skink*	*Acritoscincus duperreyi*		x	x	x	x				LC
Red-throated Skink*	*Acritoscincus platynotum*	x	x	x						LC
Western Three-lined Skink	*Acritoscincus trilineatus*					x	x			LC
McCoy's Skink*	*Anepischetosia maccoyi*		x	x						LC
Short-necked Worm-skink*	*Anomalopus brevicollis*	x								LC
Speckled Worm-skink	*Anomalopus gowi*	x								LC
Two-clawed Worm-skink	*Anomalopus leuckartii*	x	x							LC
Five-clawed Worm-skink	*Anomalopus mackayi*	x	x							LC
Cape York Worm-skink	*Anomalopus pluto*	x								LC
Swanson's Worm-skink*	*Anomalopus swansoni*		x							LC
Verreaux's Skink*	*Anomalopus verreauxii*	x	x							LC
Lyon's Grassland Striped Skink	*Austroablepharus barrylyoni*	x								CR
Kinghorn's Grassland Striped Skink	*Austroablepharus kinghorni*	x	x			x				LC
Orange-tailed Grassland Striped Skink	*Austroablepharus naranjicaudus*						x	x		LC
Major Skink*	*Bellatorias frerei*	x	x							LC
Land Mullet*	*Bellatorias major*	x	x							LC
Arnhem Land Gorges Skink	*Bellatorias obiri*							x		CR
Cone-eared Calyptotis	*Calyptotis lepidorostrum*	x								LC
Red-tailed Calyptotis	*Calyptotis ruficauda*		x							LC
Garden Calyptotis*	*Calyptotis scutirostrum*	x	x							LC
Broad-templed Calyptotis	*Calyptotis temporalis*	x								LC
Thornton Peak Calyptotis	*Calyptotis thorntonensis*	x								LC
Coventry's Skink*	*Carinascincus coventryi*		x	x						LC
Northern Snow Skink	*Carinascincus greeni*				x					VU
Metallic Skink*	*Carinascincus metallicus*			x	x					LC
Southern Snow Skink	*Carinascincus microlepidotus*				x					VU
Ocellated Skink*	*Carinascincus ocellatus*				x					LC
Tasmanian Mountain Skink	*Carinascincus orocryptus*				x					VU
Pedra Branca Skink	*Carinascincus palfreymani*				x					VU
Tasmanian Tree Skink	*Carinascincus pretiosus*				x					LC
Two-spined Rainbow Skink*	*Carlia amax*	x					x	x		LC
Northern Red-throated Rainbow Skink	*Carlia crypta*	x								LC
Elegant Rainbow Skink	*Carlia decora*	x								LC
Sandy Rainbow Skink	*Carlia dogare*	x								LC
Slender Rainbow Skink	*Carlia gracilis*						x	x		LC
Whitsunday Rainbow Skink	*Carlia inconnexa*	x								LC
Kimberley Islands Rainbow Skink	*Carlia insularis*						x			LC
Monsoonal Three-keeled Rainbow Skink	*Carlia isostriacantha*						x			LC
Lined Rainbow Skink*	*Carlia jarnoldae*	x								LC
Rough Brown Rainbow Skink	*Carlia johnstonei*						x			LC
Closed-litter Rainbow Skink	*Carlia longipes*	x								LC
Striped Rainbow Skink*	*Carlia munda*	x					x	x		LC
Open-litter Rainbow Skink*	*Carlia pectoralis*	x								LC
Five-carinated Rainbow Skink	*Carlia quinquecarinata*	x								DD
Blue-throated Rainbow Skink*	*Carlia rhomboidalis*	x								LC
Crevice Rainbow Skink	*Carlia rimula*	x								LC
Black-throated Rainbow Skink	*Carlia rostralis*	x								LC
Orange-flanked Rainbow Skink	*Carlia rubigo*	x								LC

Common English Name	Scientific Name	Qld	NSW	Vic	Tas	SA	WA	NT	Islands	IUCN
Southern Red-throated Rainbow Skink*	Carlia rubrigularis	x								LC
Red-sided Rainbow Skink	Carlia rufilatus						x	x		LC
Schmeltz's Rainbow Skink	Carlia schmeltzii	x								LC
Robust Rainbow Skink	Carlia sexdentata	x						x		LC
Storr's Rainbow Skink	Carlia storri	x								LC
Southern Rainbow Skink	Carlia tetradactyla	x	x	x						LC
Desert Rainbow Skink	Carlia triacantha	x				x	x	x		LC
Tussock Rainbow Skink	Carlia vivax	x	x							LC
Cape Melville Rainbow Skink	Carlia wundalthini	x								VU
Limbless Snake-tooth Skink*	Coeranoscincus frontalis	x								LC
Three-toed Snake-tooth Skink*	Coeranoscincus reticulatus	x	x							LC
Satinay Sand Skink*	Coggeria naufragus	x								LC
Lemon-barred Forest-skink*	Concinnia amplus	x								LC
Northern Bar-sided Skink*	Concinnia brachysoma	x								LC
Bartle Frere Bar-sided Skink	Concinnia frerei	x								VU
Martin's Skink*	Concinnia martini	x	x							LC
Stout Bar-sided Skink	Concinnia sokosoma	x								LC
Bar-sided Skink*	Concinnia tenuis	x	x							LC
Yellow-blotched Forest-skink	Concinnia tigrina	x								LC
Adam's Snake-eyed Skink	Cryptoblepharus adamsi	x								LC
Inland Snake-eyed Skink*	Cryptoblepharus australis	x	x			x	x	x		LC
Buchanan's Snake-eyed Skink*	Cryptoblepharus buchanani						x			LC
Swanson's Snake-eyed Skink	Cryptoblepharus cygnatus							x		LC
Dapple Snake-eyed Skink	Cryptoblepharus daedalos							x		LC
Christmas Island Blue-tailed Shining Skink*	Cryptoblepharus egeriae								x	EX
Noble Snake-eyed Skink	Cryptoblepharus exochus						x	x		LC
Fuhn's Snake-eyed Skink	Cryptoblepharus fuhni	x								LC
Arafura Snake-eyed Skink	Cryptoblepharus gurrmul							x		DD
Juno's Snake-eyed Skink	Cryptoblepharus juno						x	x		LC
Coastal Snake-eyed Skink*	Cryptoblepharus litoralis litoralis	x						x		LC
Blotched Snake-eyed Skink	Cryptoblepharus megastictus						x			LC
Merten's Snake-eyed Skink	Cryptoblepharus mertensi							x		LC
Metallic Snake-eyed Skink	Cryptoblepharus metallicus	x					x	x		LC
Pale Snake-eyed Skink	Cryptoblepharus ochrus					x				LC
Ragged Snake-eyed Skink*	Cryptoblepharus pannosus	x	x	x		x		x		LC
Peron's Snake-eyed Skink	Cryptoblepharus plagiocephalus						x			LC
Elegant Snake-eyed Skink*	Cryptoblepharus pulcher pulcher	x	x			x	x			LC
Tawny Snake-eyed Skink	Cryptoblepharus ruber						x	x		LC
Pygmy Snake-eyed Skink	Cryptoblepharus tytthos						x			LC
Russet Snake-eyed Skink	Cryptoblepharus ustulatus						x			LC
Striped Snake-eyed Skink	Cryptoblepharus virgatus	x								LC
Spangled Snake-eyed Skink	Cryptoblepharus wulbu							x		LC
Agile Snake-eyed Skink*	Cryptoblepharus zoticus	x						x		LC
Mitchell Grass Ctenotus	Ctenotus agrestis	x								LC
Lively Ctenotus	Ctenotus alacer	x				x	x	x		LC
Plain-backed Sandplain Ctenotus	Ctenotus alleni						x			NT
Brown-blazed Wedge-snouted Ctenotus	Ctenotus allotropis	x	x							LC
North-west Coastal Ctenotus	Ctenotus angusticeps						x			LC
Arcane Ctenotus	Ctenotus arcanus	x	x							LC
Ariadna's Ctenotus*	Ctenotus ariadnae	x				x	x	x		LC
Jabiluka Ctenotus	Ctenotus arnhemensis							x		LC
Ashy Down's Ctenotus	Ctenotus astarte	x				x				LC
Arnhem Striped Ctenotus	Ctenotus astictus							x		LC
Southern Spinifex Ctenotus	Ctenotus atlas		x			x	x			LC
West Coast Long-tailed Ctenotus	Ctenotus australis						x			LC

Common English Name	Scientific Name	Qld	NSW	Vic	Tas	SA	WA	NT	Islands	IUCN
Northern Ctenotus	Ctenotus borealis							x		LC
Short-clawed Ctenotus	Ctenotus brachyonyx	x	x	x		x				LC
Short-footed Ctenotus	Ctenotus brevipes	x								LC
Brook's Wedge-snouted Ctenotus	Ctenotus brooksi					x	x	x		LC
Plain-backed Kimberley Ctenotus	Ctenotus burbidgei						x			LC
Blue-tailed Ctenotus	Ctenotus calurus					x	x	x		LC
Capricorn Ctenotus	Ctenotus capricorni	x								LC
Chain-striped Heath Ctenotus	Ctenotus catenifer						x			LC
Brown-backed Ctenotus*	Ctenotus coggeri							x		LC
Collett's Ctenotus	Ctenotus colletti						x			LC
Ten-lined Ctenotus	Ctenotus decaneurus decaneurus	x					x	x		LC
Darling Range Heath Ctenotus	Ctenotus delli						x			LC
Pilbara Striped Ctenotus	Ctenotus duricola						x			LC
Narrow-lined Ctenotus	Ctenotus dux	x				x	x	x		LC
Brown-tailed Finesnout Ctentous	Ctenotus ehmanni						x			LC
Port Essington Ctenotus	Ctenotus essingtonii							x		LC
Bight Wedge-snouted Ctenotus	Ctenotus euclae					x	x			LC
Brown-backed Yellow-lined Ctenotus	Ctenotus eurydice	x	x							LC
Black-backed Yellow-lined Ctenotus	Ctenotus eutaenius	x								LC
West Coast Ctenotus*	Ctenotus fallens						x			LC
Kakadu Ctenotus	Ctenotus gagudju							x		LC
Jewelled Sandplain Ctenotus	Ctenotus gemmula						x			LC
Grand Ctenotus, Giant Desert Ctenotus	Ctenotus grandis grandis					x	x	x		LC
Greer's Ctenotus	Ctenotus greeri					x	x	x		LC
Chained Ctenotus	Ctenotus halysis						x			LC
Nimble Ctenotus	Ctenotus hanloni					x	x	x		LC
Stout Ctenotus	Ctenotus hebetior hebetior	x								LC
Dusky Ctenotus	Ctenotus helenae	x				x	x	x		LC
Hill's Ctenotus	Ctenotus hilli							x		LC
North West Cape Ctenotus	Ctenotus iapetus						x			LC
South-western Odd-striped Ctenotus	Ctenotus impar						x			LC
Ingram's Ctenotus	Ctenotus ingrami	x	x							LC
Plain Ctenotus	Ctenotus inornatus	x					x	x		LC
Blacksoil Ctenotus	Ctenotus joanae	x				x		x		LC
Alligator Rivers Ctenotus	Ctenotus kurnbudj							x		EN
Red-legged Ctenotus*	Ctenotus labillardieri						x			LC
Lancelin Island Ctenotus	Ctenotus lancelini						x			CR
Gravelly-soil Ctenotus*	Ctenotus lateralis	x						x		LC
Centralian Coppertail	Ctenotus leae	x				x	x	x		LC
Common Desert Ctenotus*	Ctenotus leonhardii	x	x			x	x	x		LC
Maryan's Ctenotus	Ctenotus maryani						x			LC
Whiptail Ctenotus	Ctenotus mastigura						x			LC
Median-striped Ctenotus	Ctenotus mesotes						x			LC
Military Ctenotus	Ctenotus militaris						x	x		LC
Checker-sided Ctenotus	Ctenotus mimetes						x			LC
Atherton Ctenotus	Ctenotus monticola	x								LC
Long-snouted Ctenotus	Ctenotus nasutus					x	x	x		LC
Pin-striped Finesnout Ctenotus	Ctenotus nigrilineatus						x			LC
Cooktown Ctenotus	Ctenotus nullum	x								LC
Saltbush Ctenotus	Ctenotus olympicus	x	x	x		x		x		LC
Coastal Plains Ctenotus	Ctenotus ora						x			VU
Eastern Spotted Ctenotus*	Ctenotus orientalis		x	x		x	x			LC
Western Pilbara Lined Ctenotus	Ctenotus pallasotus						x			NA
Pale-backed Ctenotus	Ctenotus pallescens						x	x		LC

Common English Name	Scientific Name	Qld	NSW	Vic	Tas	SA	WA	NT	Islands	IUCN
Leopard Ctenotus?*	Ctenotus pantherinus pantherinus	x	x			x	x	x		LC
Pianka's Ctenotus	Ctenotus piankai	x				x	x	x		LC
Pretty Ctenotus*	Ctenotus pulchellus	x						x		LC
Fourteen-lined Ctenotus	Ctenotus quattuordecimlineatus					x	x	x		LC
Quinkan Ctentous	Ctenotus quinkan	x								LC
Arnhem Land Ctenotus	Ctenotus quirinus							x		LC
Cape Heath Ctenotus	Ctenotus rawlinsoni	x								LC
Royal Ctenotus*	Ctenotus regius	x	x	x		x	x	x		LC
Kimberley Lined Ctenotus	Ctenotus rhabdotus						x	x		LC
Crack-dwelling Ctenotus	Ctenotus rimacola						x	x		LC
Eastern Striped Ctenotus*	Ctenotus robustus	x	x	x		x	x	x		LC
Beaded Ctenotus	Ctenotus rosarium	x								LC
Ruddy Ctenotus	Ctenotus rubicundus						x			LC
Rufous Finesnout Ctenotus	Ctenotus rufescens						x			LC
Rusty-shouldered Ctenotus	Ctenotus rutilans						x			LC
Rock Ctenotus, Stony-soil Ctenotus	Ctenotus saxatilis	x					x	x		LC
Spotted Black-soil Ctenotus	Ctenotus schevilli	x								LC
Barred Wedge-snouted Ctenotus	Ctenotus schomburgkii	x	x	x		x	x	x		LC
Gibber Ctenotus	Ctenotus septenarius	x				x	x	x		LC
Gravel Downs Ctenotus	Ctenotus serotinus	x								CR
North-western Sandy-loam Ctenotus	Ctenotus serventyi						x			LC
Stern Rock Ctenotus	Ctenotus severus						x			LC
Spalding's Ctenotus	Ctenotus spaldingi	x					x	x		LC
Storr's Ctenotus	Ctenotus storri							x		LC
Eastern-barred Wedge-snouted Ctenotus	Ctenotus strauchii	x	x			x		x		LC
Stripe-headed Ctenotus	Ctenotus striaticeps	x						x		LC
Stuart's Ctenotus	Ctenotus stuarti							x		EN
Sharp-browed Ctenotus	Ctenotus superciliaris						x	x		LC
Eyrean Ctenotus	Ctenotus taeniatus	x	x	x		x		x		LC
Copper-tailed Skink*	Ctenotus taeniolatus	x	x	x						LC
Tanami Ctenotus	Ctenotus tanamiensis						x	x		LC
Kimberley Wedge-snouted Ctenotus	Ctenotus tantillus						x	x		LC
Hinchinbrook Ctenotus	Ctenotus terrareginae	x								LC
Spotted Ctenotus, Western Spotted Ctenotus	Ctenotus uber uber						x			LC
Uneven-striped Ctenotus	Ctenotus vagus						x			DD
Scant-striped Ctenotus	Ctenotus vertebralis							x		LC
Wide-striped Sandplain Ctenotus	Ctenotus xenopleura						x			LC
Shark Bay Ctenotus	Ctenotus youngsoni						x			LC
Hamelin Pool Ctenotus	Ctenotus zastictus						x			NT
Little Zebra Ctenotus	Ctenotus zebrilla	x								LC
Gilled Slender Blue-tongue	Cyclodomorphus branchialis						x			NT
Tasmanian She-oak Skink*	Cyclodomorphus casuarinae				x					LC
Western Slender Blue-tongue*	Cyclodomorphus celatus						x			LC
Pink-tongued Skink*	Cyclodomorphus gerrardii	x	x							LC
Giant Slender Blue-tongue	Cyclodomorphus maximus						x			LC
Spinifex Slender Blue-tongue*	Cyclodomorphus melanops melanops	x	x			x	x	x		LC
Mainland She-oak Skink*	Cyclodomorphus michaeli		x	x						LC
Alpine She-oak Skink	Cyclodomorphus praealtus		x	x						EN
Saltbush Slender Blue-tongue	Cyclodomorphus venustus	x	x			x				LC
Cunningham's Skink*	Egernia cunninghami	x	x	x		x				LC
Western Pilbara Spiny-tailed Skink	Egernia cygnitos						x			LC
Southern Pygmy Spiny-tailed Skink*	Egernia depressa						x			LC

Common English Name	Scientific Name	Qld	NSW	Vic	Tas	SA	WA	NT	Islands	IUCN
Kimberley Crevice-skink	Egernia douglasi						x			DD
Central Pygmy Spiny-tailed Skink	Egernia eos						x	x		LC
Eastern Pilbara Spiny-tailed Skink*	Egernia epsisolus						x			LC
Goldfields Crevice-skink	Egernia formosa						x			LC
Hosmer's Skink*	Egernia hosmeri	x						x		LC
King's Skink*	Egernia kingii						x			LC
Eastern Crevice Skink*	Egernia mcpheei	x	x							LC
South-western Crevice-skink*	Egernia napoleonis						x			LC
Pilbara Crevice-skink	Egernia pilbarensis						x			LC
Richard's Crevice-skink	Egernia richardi					x	x			LC
Mount Kaputar Skink	Egernia roomi		x							NA
Yakka Skink*	Egernia rugosa	x								LC
Black Rock Skink*	Egernia saxatilis saxatilis		x	x						LC
Gidgee Skink?*	Egernia stokesii stokesii	x	x?			x	x	x		LC
Tree Skink*	Egernia striolata	x	x	x		x				LC
Mangrove Skink	Emoia atrocostata	x							x	LC
Long-tailed Skink	Emoia longicauda	x								LC
Christmas Island Forest Skink	Emoia nativitatus								x	EX
Brown-sided Bar-lipped Skink	Eremiascincusus brongersmai						x			LC
Orange-sided Bar-lipped Skink	Eremiascincus douglasi							x		LC
Eastern Narrow-banded Sand-swimmer*	Eremiascincus fasciolatus	x								LC
Northern Narrow-banded Skink	Eremiascincus intermedius						x	x		LC
Northern Bar-lipped Skink	Eremiascincus isolepis	x					x	x		LC
Mosaic Desert Skink	Eremiascincus musivus						x			LC
Western Narrow-banded Skink	Eremiascincus pallidus					x	x	x		LC
Lowlands Bar-lipped Skink	Eremiascincus pardalis	x								LC
Ghost Skink	Eremiascincus phantasmus	x	x			x		x		LC
Broad-banded Sand-swimmer*	Eremiascincus richardsonii	x	x			x	x	x		LC
Rusty Skink	Eremiascincus rubiginosus						x			LC
Elf Skink*	Eroticoscincus graciloides	x								LC
Brown Sheen-skink*	Eugongylus rufescens	x								LC
Yellow-bellied Water Skink*	Eulamprus heatwolei		x	x		x				LC
Alpine Water Skink	Eulamprus kosciuskoi	x	x	x						LC
Blue Mountains Water Skink*	Eulamprus leuraensis		x							EN
Eastern Water Skink*	Eulamprus quoyii	x	x	x		x				LC
Southern Water Skink*	Eulamprus tympanum tympanum		x	x		x				LC
Mount Elliot Mulch-skink	Glaphyromorphus clandestinus	x								LC
Slender Mulch-skink	Glaphyromorphus cracens	x								LC
Northern Mulch-skink	Glaphyromorphus crassicaudus	x								LC
Top End Mulch-skink	Glaphyromorphus darwiniensis							x		LC
Brown-tailed Bar-lipped Skink	Glaphyromorphus fuscicaudis	x								LC
Atherton Tableland Mulch-skink*	Glaphyromorphus mjobergi	x								LC
Black-tailed Bar-lipped Skink	Glaphyromorphus nigricaudis	x						x		LC
McIlwraith Bar-lipped Skink	Glaphyromorphus nyanchupinta	x								LC
Cape Melville Bar-lipped Skink	Glaphyromorphus othelarrni	x								LC
Dwarf Mulch-skink	Glaphyromorphus pumilus	x								LC
Fine-spotted Mulch-skink*	Glaphyromorphus punctulatus	x								LC
Prickly Forest Skink*	Gnypetoscincus queenslandiae	x								LC
Rainforest Cool-skink*	Harrisoniascincus zia	x	x							LC
Three-toed Earless Skink*	Hemiergis decresiensis decresiensis			x		x				LC
South-western Mulch-skink	Hemiergis gracilipes						x			LC
Southwestern Earless Skink	Hemiergis initialis initialis					x	x			LC
Rusty Earless Skink	Hemiergis millewae		x	x		x	x			LC
Peron's Earless Skink	Hemiergis peronii peronii			x		x	x			LC

Common English Name	Scientific Name	Qld	NSW	Vic	Tas	SA	WA	NT	Islands	IUCN
Two-toed Earless Skink*	Hemiergis quadrilineatum						x			LC
Eastern Earless Skink	Hemiergis talbingoensis talbingoensis	x	x	x						LC
Murray's Skink*	Karma murrayi	x	x							LC
Tryon's Skink	Karma tryoni	x	x							LC
Diamond-shielded Sunskink	Lampropholis adonis	x								LC
Friendly Sunskink	Lampropholis amicula	x	x							LC
Southern Montane Sunskink	Lampropholis bellendenkerensis	x								VU
Montane Sunskink, Barrington Sunskink	Lampropholis caligula		x							LC
Rainforest Sunskink*	Lampropholis coggeri	x								LC
Bunya Mountains Sunskink	Lampropholis colossus	x								DD
Couper's Sunskink	Lampropholis couperi	x								LC
Garden Skink*	Lampropholis delicata	x	x	x	x	x				LC
Southern Sunskink	Lampropholis elliotensis	x								VU
Elongate Sunskink	Lampropholis elongata		x							DD
Grass Skink*	Lampropholis guichenoti	x	x	x		x				LC
Saxicoline Sunskink	Lampropholis mirabilis	x								LC
Robert's Sunskink	Lampropholis robertsi	x								LC
Southern Rainforest Sunskink	Lampropholis similis	x								LC
Yellow-tailed Slider	Lerista aericeps	x	x			x		x		LC
Retro Slider	Lerista allanae	x								CR
Bulleringa Fine-lined Slider	Lerista alia	x								NA
Cape Range Slider	Lerista allochira						x			LC
Limbless Fine-lined Slider	Lerista ameles	x								EN
Fortescue Three-toed Slider	Lerista amicorum						x			LC
Olkola Slider	Lerista anyara	x								NA
Dampierland Limbless Slider	Lerista apoda						x			LC
Bight Slider	Lerista arenicola					x	x			LC
Kalbarri Robust Slider	Lerista axillaris						x			DD
Bayne's Slider	Lerista baynesi						x			LC
Western Two-toed Slider*	Lerista bipes	x				x	x	x		LC
Northern Slider	Lerista borealis						x	x		LC
Bougainville's Slider*	Lerista bougainvillii		x	x	x	x				LC
Bungle Bungle Robust Slider	Lerista bunglebungle						x			DD
Carpentaria Slider	Lerista carpentariae							x		LC
Lyre-patterned Slider	Lerista chordae	x								VU
Bold-striped Four-toed Slider	Lerista christinae						x			VU
Vine-thicket Fine-lined Slider	Lerista cinerea	x								NT
Sharp-blazed Three-toed Slider	Lerista clara						x			LC
Colliver's Slider	Lerista colliveri	x								NT
Blinking Broad-striped Slider	Lerista connivens						x			LC
Great Desert Slider	Lerista desertorum	x				x	x	x		LC
South-western Four-toed Slider	Lerista distinguenda					x	x			LC
Southern Four-toed Slider	Lerista dorsalis					x	x			LC
Myall Slider*	Lerista edwardsae					x				LC
Elegant Slider	Lerista elegans						x			LC
Woomera Slider	Lerista elongata					x				LC
Noonbah Robust Slider	Lerista emmotti	x	x			x				LC
Meekatharra Slider	Lerista eupoda						x			LC
Pilbara Flame-tailed Slider	Lerista flammicauda						x			LC
Eastern Mulch-slider	Lerista fragilis	x								LC
Centralian Four-toed Slider	Lerista frosti							x		LC
Gascoyne Broad-striped Slider	Lerista gascoynensis						x			LC
Bold-striped Robust Slider	Lerista gerrardii						x			LC
South-eastern Kimberley Sand-slider	Lerista greeri						x	x		LC
Stout Sand Slider	Lerista griffini						x	x		LC
Gnaraloo Three-toed Slider	Lerista haroldi						x			EN
Hobson's Fine-lined Slider	Lerista hobsoni	x								NT

Common English Name	Scientific Name	Qld	NSW	Vic	Tas	SA	WA	NT	Islands	IUCN
Humphries Worm Slider	Lerista humphriesi						x			LC
McIvor River Slider	Lerista ingrami	x								DD
Robust Worm-slider	Lerista ips						x	x		LC
Jackson's Three-toed Slider	Lerista jacksoni						x			LC
Kalumburu Kimberley Slider	Lerista kalumburu						x			LC
Karl Schmidt's Slider	Lerista karlschmidti	x						x		LC
Shark Bay Broad-striped Slider	Lerista kendricki						x			LC
Kennedy Range Broad-striped Slider	Lerista kennedyensis						x			LC
King's Three-toed Slider	Lerista kingi						x			LC
Eastern Two-toed Slider	Lerista labialis	x	x			x	x	x		LC
Perth Lined Slider	Lerista lineata						x			EN
Line-spotted Robust Slider	Lerista lineopunctulata						x			LC
Unpatterned Robust Slider	Lerista macropisthopus macropisthopus						x			LC
Micro Three-toed Slider	Lerista micra						x			LC
South-western Slider	Lerista microtis microtis					x	x			LC
Northern Dotted-line Robust Slider	Lerista miopus						x			LC
Mueller's Three-toed Slider	Lerista muelleri						x			LC
Pilbara Robust Slider	Lerista neander						x			LC
Nevin's Three-toed Slider	Lerista nevinae						x			CR
Inland Broad-striped Slider	Lerista nichollsi						x			LC
Hidden Three-toed Slider	Lerista occulta						x			LC
Onslow Broad-striped Slider	Lerista onsloviana						x			LC
North-eastern Slider	Lerista orientalis	x					x	x		LC
Chillagoe Fine-lined Slider	Lerista parameles	x								NA
Pale Broad-striped Slider	Lerista petersoni						x			LC
Southern Robust Slider	Lerista picturata						x			LC
Keeled Slider	Lerista planiventralis planiventralis						x			LC
Yampi Sand-slider	Lerista praefrontalis						x			DD
West Coast Worm-slider	Lerista praepedita						x			LC
Eastern Robust Slider*	Lerista punctatovittata	x	x	x		x				LC
Great Victoria Desert Slider	Lerista puncticauda						x			EN
Dark-streaked Slider	Lerista quadrivincula						x			DD
Dusky Slider	Lerista robusta						x			DD
Rochford Slider	Lerista rochfordensis	x								VU
Rolfe's Three-toed Slider	Lerista rolfei						x			LC
Dampierland Slider	Lerista separanda						x			LC
Fitzroy Sand-slider	Lerista simillima						x			LC
Musgrave Slider	Lerista speciosa					x				LC
Spotted Broad-striped Slider	Lerista stictopleura						x			LC
Chillagoe Fine-lined Slider	Lerista storri	x								NT
Single-toed Slider	Lerista stylis							x		LC
Ribbon Slider	Lerista taeniata					x	x	x		LC
Southern Three-toed Slider	Lerista terdigitata					x				LC
Dwarf Three-toed Slider*	Lerista timida	x	x	x		x	x	x		LC
Dark-backed Mulch Slider	Lerista tridactyla						x			LC
Slender Broad-striped Slider	Lerista uniduo						x			LC
Leaden-bellied Fine-lined Slider	Lerista vanderduysi	x								NT
Variable-striped Robust Slider	Lerista varia						x			LC
Powerful Three-toed Slider	Lerista verhmens						x			LC
Slender Worm-slider	Lerista vermicularis						x			LC
Ravensthorpe Range Slider	Lerista viduata						x			DD
Side-striped Fine-lined Slider	Lerista vittata	x								CR
Walker's Slider	Lerista walkeri						x			LC
Wilkin's Slider	Lerista wilkinsi	x								LC

Common English Name	Scientific Name	Qld	NSW	Vic	Tas	SA	WA	NT	Islands	IUCN
Yellow-tailed Plain Slider	*Lerista xanthura*						x			LC
Yuna Broad-striped Slider	*Lerista yuna*						x			NT
Pilbara Blue-tailed Slider	*Lerista zietzi*						x			LC
Wide-striped Four-toed Slider	*Lerista zonulata*	x								LC
Bamboo Range Rock Skink	*Liburnascincus artemis*	x								LC
Coen Rainbow Skink	*Liburnascincus coensis*	x								LC
Outcrop Rock Skink*	*Liburnascincus mundivensis*	x								LC
Black Mountain Rock Skink*	*Liburnascincus scirtetis*	x								LC
Guthega Skink*	*Liopholis guthega*		x	x						
Desert Skink *	*Liopholis inornata*	x	x	x		x	x	x		LC
Great Desert Skink, Tjakura*	*Liopholis kintorei*					x	x	x		VU
Masked Rock Skink*	*Liopholis margaretae*					x		x		LC
Eastern Ranges Rock Skink	*Liopholis modesta*	x	x							LC
Montane Rock Skink	*Liopholis montana*		x	x						NT
Bull Skink, Heath Skink	*Liopholis multiscutata*			x		x	x			LC
Flinders Ranges Rock Skink	*Liopholis personata*					x				LC
South-western Rock Skink	*Liopholis pulchra pulchra*						x			LC
Slater's Skink	*Liopholis slateri slateri*					x		x		VU
Night Skink*	*Liopholis striata*					x	x	x		LC
White's Skink*	*Liopholis whitii*	x	x	x	x	x				LC
Swamp Skink*	*Lissolepis coventryi*		x	x		x				EN
Mourning Skink*	*Lissolepis luctuosa*						x			LC
Secretive Litter-skink	*Lygisaurus abscondita*	x								DD
Large-disced Litter-skink	*Lygisaurus aeratus*	x								LC
Tree-base Litter-skink*	*Lygisaurus foliorum*	x	x							LC
Rainforest Edge Litter-skink	*Lygisaurus laevis*	x								LC
Translucent Litter-skink	*Lygisaurus macfarlani*	x						x		LC
Red-tailed Litter-skink	*Lygisaurus malleolus*	x								LC
Fire-tailed Litter-skink	*Lygisaurus parrhasius*	x								LC
Chillagoe Litter-skink	*Lygisaurus rococo*	x								LC
Eastern Cape Litter-skink	*Lygisaurus sesbrauna*	x								LC
Tanner's Litter-skink	*Lygisaurus tanneri*	x								LC
Sun-loving Litter-skink	*Lygisaurus zuma*	x								LC
Christmas Island Skink	*Lygosoma bowringi*								x	NA
Orange-speckled Forest Skink*	*Magmellia luteilateralis*	x								DD
Top-end Dwarf Skink	*Menetia alanae*							x		LC
Jabiluka Dwarf Skink	*Menetia concinna*							x		LC
Common Dwarf Skink*	*Menetia greyii*	x	x	x		x	x	x		LC
Main's Dwarf Skink	*Menetia maini*	x					x	x		LC
Western Dwarf Skink	*Menetia surda surda*						x			LC
Saltbush Morethia*	*Morethia adelaidensis*	x	x	x		x	x	x		LC
Boulenger's Morethia*	*Morethia boulengeri*	x	x	x		x	x	x		LC
Woodland Dark-flecked Morethia	*Morethia butleri*					x	x	x		LC
West Coast Pale-flecked Morethia	*Morethia lineoocellata*						x			LC
Shrubland Pale-flecked Morethia	*Morethia obscura*		x	x		x	x			LC
Lined Fire-tail Skink*	*Morethia ruficauda ruficauda*	x				x	x	x		LC
Northern Fire-tailed Skink	*Morethia storri*						x	x		LC
Eastern Fire-tailed Skink	*Morethia taeniopleura*	x								LC
Nangur Skink*	*Nangura spinosa*	x								EN
Western Striped Snake-eyed Skink	*Notoscincus butleri*						x			LC
Desert Glossy Skink*	*Notoscincus ornatus ornatus*	x				x	x	x		LC
Lord Howe Island Skink	*Oligosoma lichenigera*								x	VU
Cooloola Snake-skink	*Ophioscincus cooloolensis*	x								LC
Yolk-bellied Snake-skink*	*Ophioscincus ophioscincus*	x								LC
Short-limbed Snake-skink	*Ophioscincus truncatus*	x	x							LC
Spinifex Snake-eyed Skink	*Proablepharus reginae*					x	x	x		LC
Slender Snake-eyed Skink	*Proablepharus tenuis*	x					x	x		LC
Bight Coast Skink	*Pseudemoia baudini*					x	x			LC
Alpine Bog Skink	*Pseudemoia cryodroma*			x						EN

Common English Name	Scientific Name	Qld	NSW	Vic	Tas	SA	WA	NT	Islands	IUCN
Southern Grass Skink*	Pseudemoia entrecasteauxii		x	x	x	x				LC
Tussock Skink*	Pseudemoia pagenstecheri		x	x	x					LC
Glossy Grass Skink	Pseudemoia rawlinsoni		x	x	x	x				DD
Spencer's Skink*	Pseudemoia spenceri		x	x						LC
Koshland's Dwarf Skink	Pygmaeascincus koschlandae	x								LC
Magnetic Island Dwarf Skink	Pygmaeascincus sadlieri	x								LC
Low's Dwarf Skink	Pygmaeascincus timlowi	x								LC
Three-toed Skink*	Saiphos equalis	x	x							LC
Southern Wet Tropics Shadeskink*	Saproscincus basiliscus	x								LC
Challenger's Shadeskink	Saproscincus challengeri	x	x							LC
Czechura's Shadeskink	Saproscincus czechurai	x								LC
Eungella Shadeskink*	Saproscincus eungellensis	x								DD
Hannah's Shadeskink	Saproscincus hannahae	x								LC
Northern Wet Tropics Shadeskink	Saproscincus lewisi	x								LC
Weasel Skink*	Saproscincus mustelinus		x	x						LC
Coastal Shadeskink	Saproscincus orianus	x	x							LC
Rose's Shadeskink*	Saproscincus rosei	x	x							LC
Cape Melville Shadeskink	Saproscincus saltus	x								VU
Gully Skink	Saproscincus spectabilis	x	x							LC
Four-fingered Shadeskink*	Saproscincus tetradactylus	x								LC
Bartle Frere Skink	Techmarscincus jigurru	x								VU
Pygmy Blue-tongue*	Tiliqua adelaidensis					x				EN
Centralian Blue-tongue*	Tiliqua multifasciata	x	x			x	x	x		LC
Blotched Blue-tongue*	Tiliqua nigrolutea		x	x	x					LC
Western Blue-tongue*	Tiliqua occipitalis		x	x		x	x	x		LC
Shingleback Lizard*	Tiliqua rugosa rugosa	x	x	x		x	x			LC
Common Blue-tongue, Eastern Blue-tongue	Tiliqua scincoides scincoides	x	x	x		x	x	x		LC
Dragons (Agamidae)										
Burn's Dragon*	Amphibolurus burnsi	x	x			x				LC
Centralian Tree Dragon	Amphibolurus centralis	x						x		
Jacky Lizard*	Amphibolurus muricatus	x	x	x		x				LC
Mallee Tree Dragon	Amphibolurus norrisi			x		x	x			LC
Mulga Dragon*	Caimanops amphiboluroides						x			LC
Chameleon Dragon*	Chelosania brunnea	x					x	x		DD
Frilled Dragon*	Chlamydosaurus kingii	x					x	x		LC
Hidden Dragon	Cryptagama aurita						x	x		DD
Western Heath Dragon	Ctenophorus adelaidensis						x			LC
Shark Bay Heath Dragon	Ctenophorus butleri						x			LC
Western Ring-tailed Dragon	Ctenophorus caudicinctus						x			LC
Southern Heath Dragon	Ctenophorus chapmani					x	x			LC
Black-collared Dragon	Ctenophorus clayi					x	x	x		LC
Crested Dragon, Bicycle Lizard*	Ctenophorus cristatus					x	x			LC
Tawny Dragon*	Ctenophorus decresii					x				LC
Long-tailed Sand Dragon	Ctenophorus femoralis						x			LC
Peninsula Dragon*	Ctenophorus fionni					x				LC
Mallee Military Dragon*	Ctenophorus fordi	x	x	x		x	x			LC
Gibber Dragon*	Ctenophorus gibba					x				LC
Graaf's Ring-tailed Dragon	Ctenophorus graafi						x	x		LC
Goldfields Ring-tailed Dragon	Ctenophorus infans						x			LC
Central Military Dragon*	Ctenophorus isolepis isolepis	x				x	x	x		LC
Spotted Military Dragon	Ctenophorus maculatus maculatus					x	x			LC
Lake Eyre Dragon*	Ctenophorus maculosus					x				LC
McKenzie's Dragon	Ctenophorus mckenziei					x	x			LC
Barrier Range Dragon	Ctenophorus mirrityana		x							NT
Lake Disappointment Dragon	Ctenophorus nguyarna						x			VU
Central Netted Dragon*	Ctenophorus nuchalis	x	x			x	x	x		LC

Common English Name	Scientific Name	Qld	NSW	Vic	Tas	SA	WA	NT	Islands	IUCN
Ornate Dragon*	Ctenophorus ornatus						X			LC
North-western Heath Dragon	Ctenophorus parviceps						X			LC
Painted Dragon*	Ctenophorus pictus	X	X	X		X	X	X		LC
Western Netted Dragon*	Ctenophorus reticulatus					X	X	X		LC
Rufus Sand Dragon	Ctenophorus rubens						X			LC
Rusty Dragon	Ctenophorus rufescens					X	X	X		LC
Claypan Dragon*	Ctenophorus salinarum					X	X			LC
Lozenge-marked Dragon*	Ctenophorus scutulatus						X			LC
Slater's Ring-tailed Dragon*	Ctenophorus slateri	X					X	X		LC
Eastern Mallee Sand Dragon	Ctenophorus spinodomus		X							NA
Ochre Dragon	Ctenophorus tjantjalka					X				LC
Red-barred Dragon	Ctenophorus vadnappa					X				LC
Yinnietharra Rock Dragon	Ctenophorus yinnietharra						X			LC
Carnarvon Dragon	Diporiphora adductus						X			LC
White-lipped Two-lined Dragon	Diporiphora albilabris						X	X		LC
Amelia's Spinifex Dragon	Diporiphora ameliae	X								LC
Tommy Roundhead*	Diporiphora australis	X	X							LC
Robust Two-lined Dragon	Diporiphora bennettii						X	X		LC
Two-lined Dragon	Diporiphora bilineata	X					X	X		LC
Gulf Two-lined Dragon	Diporiphora carpentariensis	X								LC
Crystal Creek Dragon	Diporiphora convergens						X			DD
Gracile Two-lined Dragon	Diporiphora gracilis						X			NA
Granulated Two-lined Dragon*	Diporiphora granulifera	X						X		LC
Black Throated Two-lined Dragon*	Diporiphora jugularis	X								LC
Lally's Two-lined Dragon*	Diporiphora lalliae						X	X		LC
Linga Dragon	Diporiphora linga					X	X			LC
Yellow-sided Two-lined Dragon*	Diporiphora magna	X					X	X		LC
Margaret's Two-lined Dragon	Diporiphora margaretae						X			NA
Common Nobbi Dragon*	Diporiphora nobbi	X	X	X		X				LC
Pale two-Pored Dragon	Diporiphora pallidus						X			NA
Grey-striped Western Desert Dragon	Diporiphora paraconvergens					X	X	X		LC
Kimberley Two-pored Rock Dragon	Diporiphora perplexa						X	X		NA
Black-throated Nobbi Dragon	Diporiphora phaeospinosa	X								LC
Pindan Dragon	Diporiphora pindan						X			LC
Red-rumped Two-lined Dragon*	Diporiphora reginae					X	X			LC
Northern Two-pored Rock Dragon	Diporiphora sobria	X					X	X		NA
Superb Dragon*	Diporiphora superba						X			LC
Southern Pilbara Tree Dragon	Diporiphora valens						X			LC
Northern Pilbara Tree Dragon	Diporiphora vescus						X			VU
Canegrass Dragon*	Diporiphora winneckei	X	X			X		X		LC
Long-nosed Dragon*	Gowidon longirostris	X				X	X	X		LC
Gippsland Water Dragon*	Intellagama lesueurii howitti		X	X						LC
Eastern Water Dragon*	Intellagama lesueurii lesueurii	X	X	X						LC
Gilbert's Dragon, Ta-ta Lizard	Lophognathus gilberti	X					X	X		LC
Horner's Dragon*	Lophognathus horneri	X					X	X		NA
Boyd's Forest Dragon*	Lophosaurus boydii	X								LC
Southern Angle-headed Dragon*	Lophosaurus spinipes	X	X							LC
Thorny Devil*	Moloch horridus	X				X	X	X		LC
Eastern Bearded Dragon*	Pogona barbata	X	X	X		X				LC
Downs Bearded Dragon*	Pogona henrylawsoni	X								LC
Small-scaled Bearded Dragon	Pogona microlepidota						X			LC
Western Bearded Dragon*	Pogona minor minor					X	X	X		LC
Nullarbor Bearded Dragon	Pogona nullarbor					X	X			LC
Central Bearded Dragon*	Pogona vitticeps	X	X	X		X		X		LC
Mountain Heath Dragon*	Rankinia diemensis		X	X	X					LC
Swamplands Lashtail*	Tropicagama temporalis	X						X		LC
Claypan Earless Dragon	Tympanocryptis argillosa	x?	X			X				NA

169

Common English Name	Scientific Name	Qld	NSW	Vic	Tas	SA	WA	NT	Islands	IUCN
Centralian Earless Dragon	Tympanocryptis centralis					x	x	x		LC
Coastal Pebble-mimic Dragon	Tympanocryptis cephalus						x			DD
Condamine Earless Dragon*	Tympanocryptis condaminensis	x								EN
Hamersley Pebble-mimic Dragon	Tympanocryptis diabolicus						x			LC
Harlequin Earless Dragon	Tympanocryptis fictilis					x				NA
Fortescue Pebble-mimic Dragon	Tympanocryptis fortescuensis						x			LC
Gascoyne Pebble-mimic Dragon	Tympanocryptis gigas						x			LC
Nullarbor Earless Dragon	Tympanocryptis houstoni					x	x			LC
Smooth-snouted Earless Dragon*	Tympanocryptis intima	x	x			x		x		LC
Canberra Earless Dragon*	Tympanocryptis lineata		x							LC
Savannah Earless Dragon	Tympanocryptis macra						x	x		LC
Bathurst Grassland Earless Dragon	Tympanocryptis mccartneyi		x							NA
Monaro Grassland Earless Dragon	Tympanocryptis osbornei		x							NA
Five-lined Earless Dragon	Tympanocryptis pentalineata	x								DD
Lined Earless Dragon	Tympanocryptis petersi		x	x		x				NA
Grassland Earless Dragon	Tympanocryptis pinguicolla			x						EX
Goldfields Pebble-mimic Dragon	Tympanocryptis pseudopsephos						x			LC
Tennant Creek Earless Dragon	Tympanocryptis rustica							x		NA
Eyrean Earless Dragon*	Tympanocryptis tetraporophora	x	x			x		x		LC
Gawler Earless Dragon	Tympanocryptis tolleyi					x				NA
Even-scaled Earless Dragon	Tympanocryptis uniformis							x		DD
Roma Earless Dragon	Tympanocryptis wilsoni	x								EN
Monitors or Goannas (Varanidae)										
Ridge-tailed Monitor*	Varanus acanthurus acanthurus	x				x	x	x		LC
Black-spotted Spiny-tailed Monitor*	Varanus baritji							x		LC
Short-tailed Pygmy Monitor*	Varanus brevicauda	x				x	x	x		LC
Bush's Monitor*	Varanus bushi						x			LC
Stripe-tailed Monitor*	Varanus caudolineatus						x			LC
Mangrove Monitor*	Varanus chlorostigma	x						x		LC
Blue-tailed Monitor*	Varanus doreanus	x								LC
Pygmy Desert Monitor*	Varanus eremius	x				x	x	x		LC
Perentie*	Varanus giganteus	x				x	x	x		LC
Pygmy Mulga Monitor*	Varanus gilleni	x				x	x	x		LC
Kimberley Rock Monitor*	Varanus glauerti						x	x		LC
Twilight Monitor*	Varanus glebopalma	x					x	x		LC
Gould's Goanna*	Varanus gouldii gouldii	x	x	x		x	x	x		LC
Southern Pilbara Rock Monitor*	Varanus hamersleyensis						x			LC
Canopy Monitor*	Varanus keithhornei	x								LC
Long-tailed Rock Monitor*	Varanus kingorum						x	x		LC
Merten's Water Monitor*	Varanus mertensi	x					x	x		EN
Mitchell's Water Monitor*	Varanus mitchelli	x					x	x		CR
Yellow-spotted Monitor*	Varanus panoptes panoptes	x				x	x	x	x	LC
Northern Pilbara Rock Monitor*	Varanus pilbarensis						x			LC
Emerald Monitor*	Varanus prasinus	x							x	LC
Blunt-spined Monitor*	Varanus primordius							x		LC
Heath Monitor*	Varanus rosenbergi		x	x		x	x			LC
Spotted Tree Monitor*	Varanus scalaris scalaris	x					x	x		LC
Rusty Monitor*	Varanus semiremex	x								LC
Dampier Peninsula Monitor*	Varanus sparnus						x			DD
Spencer's Monitor*	Varanus spenceri	x						x		LC
Storr's Monitor*	Varanus storri storri	x					x	x		LC
Black-headed Monitor*	Varanus tristis tristis	x	x			x	x	x		LC
Lace Monitor*	Varanus varius	x	x	x		x				LC

■ FURTHER READING ■

REFERENCES

Amey, A. P., Couper, P. J. & Worthington Wilmer, J. 2019. A new species of *Lerista* Bell, 1833 (Reptilia: Scincidae) from Cape York Peninsula, Queensland, belonging to the *Lerista allanae* clade but strongly disjunct from other members of the clade. *Zootaxa* 4613 (1).

Amey, A. P., Couper, P. J. & Worthington Wilmer, J. 2019. Two new species of *Lerista* Bell, 1833 (Reptilia: Scincidae) from north Queensland populations formerly assigned to *Lerista storri* Greer, McDonald and Lawrie, 1983. *Zootaxa* 4577 (1).

Auliya, M. & Koch, A. 2020. *Visual Identification Guide for the Monitor Lizard Species of the World (Genus Varanus)*. Federal Agency for Nature Conservation, Bonn.

Brown, D. 2012. *A Guide to Australian Dragons in Captivity*. Reptile Keeper Publications: Burleigh Heads, NSW.

Brown, D. 2012. *A Guide to Australian Geckos & Pygopods in Captivity*. Reptile Keeper Publications: Burleigh Heads, NSW.

Brown, D. 2012. *A Guide to Australian Monitors in Captivity*. Reptile Keeper Publications: Burleigh Heads, NSW.

Brown, D. 2012. *A Guide to Australian Skinks in Captivity*. Reptile Keeper Publications: Burleigh Heads, NSW.

Brown, D. 2014. *A Guide to Australian Lizards in Captivity*. Reptile Keeper Publications: Burleigh Heads, NSW.

Chapple, D. G., Tingley, R., Mitchell, N. J., Macdonald, S. J., Keogh, J. S., Shea, G. M., Bowles, P., Cox, N. A. & Woinarski, J. C. Z. 2019. *The Action Plan for Australian Lizards and Snakes 2017*. CSIRO Publishing, Clayton South.

Cogger, H. 2018. *Reptiles and Amphibians of Australia* (7th edn). CSIRO Publishing, Collingwood.

Ehmann, H. 1992. *Encyclopaedia of Australian Animals – Reptiles*. Collins Angus & Robertson, Sydney, NSW.

Eidenmuller, B. & Philippen, H. 2008. *Varanoid Lizards*. Terralog, Edition Chimaira, Frankfurt.

Eipper, S. C. & Eipper, T. 2019. *A Naturalist's Guide to the Snakes of Australia*. John Beaufoy Publishing, Oxford.

Greer, A. E. 1989. *The Biology and Evolution of Australian Lizards*. Surrey Beatty and Sons, Chipping Norton, NSW.

Fry B. G., Vidal N., Norman J. A., Vonk F. J., Scheib H., Ramjan S. F. R., Kuruppu S., Fung K., Hedges A. B., Richardson M. K., Hodgson W. C., Ignjatovic V., Summerhayes R. & Kochva E. 2006. Early evolution of the venom system in lizards and snakes. *Nature* 439 584–588.

Greer, A. E. 2020. *Encyclopedia of Australian Reptiles*. Version 1. August 2020.

Kealley, L., Doughty, P., Danielle, E. & Brennan, I. B. 2020. Taxonomic assessment of two pygopod gecko subspecies from Western Australia. *Israel Journal of Ecology and Evolution*.

McCoy, M. 1980. *Reptiles of the Solomon Islands*. Wau Ecology Institute Handbook. Wau, Papua New Guinea.

Melville, J., Chaplin, K., Hipsley, C. A., Sarre, S. D., Sumner, J. & Hutchinson, M. 2019. Integrating phylogeography and high-resolution X-ray CT reveals five new cryptic species and multiple hybrid zones among Australian earless dragons. *R. Soc. open sci.* 6: 191166.

Melville, J. & Wilson, S. K. 2019. *Dragon Lizards of Australia: Evolution, Ecology & A Comprehensive Field Guide*. Museums Victoria Publishing, Melbourne, Victoria.

Oliver, P. M., Prasetya, A. M., Tedeschi, L. G., Fenker, J., Ellis, R. J., Doughty, P. & Moritz, C. 2020. Crypsis and convergence: integrative taxonomic revision of the *Gehyra australis* group (Squamata: Gekkonidae) from northern Australia. *PeerJ* 8:e7971.

Pianka, E. R., King, D. R. & King, R. A. 2004 *Varanoid Lizards of the World*. Indiana University Press, Bloomington, Indiana.

Rowland, P. & Eipper, S. C. 2018. *A Naturalist's Guide to the Dangerous Creatures of Australia*. John Beaufoy Publishing, Oxford.

Somaweera, R. 2017. *A Naturalist's Guide to the Reptiles and Amphibians of Bali*. John Beaufoy Publishing, Oxford.

Storr, G. M., Smith, L. A. & Johnstone, R. E. 1983. *Lizards of Western Australia II: Dragons and Monitors*. Western Australian Museum, Perth.

Storr, G. M., Smith, L. A. & Johnstone R. E. 1990. *Lizards of Western Australia III: Geckos and Pygopods*. Western Australian Museum, Perth.

Storr, G. M., Smith, L. A. & Johnstone, R. E. 1999. *Lizards of Western Australia I: Skinks*. Western Australian Museum, Perth.

Swan, G., Shea, G. & Sadlier, R. 2017. *A Field Guide to the Reptiles of NSW*. Reed New Holland, Sydney, NSW.

Swanson, S. 2017. *Field Guide to Australian Reptiles* (3rd edn). Pascal Press, Glebe.

The IUCN Red List of Threatened Species. www.iucnredlist.org. Downloaded on 20 March 2018.

Vincent, M. & Wilson, S. 1999. *Australian Goannas*. New Holland, Chatswood, Sydney.

Weijola, V., Vahtera, V., Koch, A. & Kraus, F. 2020. Taxonomy of Micronesian monitors (Reptilia: Squamata: *Varanus*): endemic status of new species argues for caution in pursuing eradication plans. *Royal Society of Open Science* 7:200092.

Wells, R. & Wellington, C. R. 1985. A classification of the Amphibia and Reptilia of Australia. *Australian Journal of Herpetology* (Supplementary Series) No 1: 1–64.

Wilson, S. K. 2012. *Australian Lizards – a Natural History*. CSIRO Publishing, Collingwood.

Wilson, S. K. & Swan, G. 2021. *A Complete Guide to Reptiles of Australia* (6th edn). New Holland, Chatswood, Sydney.

Wilson, S. K. 2020. *Reed Concise Guide: Lizards of Australia*. Reed New Holland, Sydney.

Websites

Atlas of Living Australia	www.ala.org.au
Australian Faunal Directory	www.biodiversity.org.au
Australian Reptile Online Database (AROD)	www.arod.com.au/arod
Australia's Wildlife	www.australiaswildlife.com
Herpmapper	www.herpmapper.org
Nature 4 You	www.wildlifedemonstrations.com

Acknowledgements

We would first like to thank our sons, Bailey and Cody. Their assistance in helping maintain many lizards and other reptiles over the years is greatly appreciated – we know your 'household chores' differed from those of your friends.

Special thanks to the following people for their assistance in making this book possible. This includes but is not limited to access to specimens in their care, assistance in the field, supply of images and thought-provoking conversations. Thank you Andrew Amey, Joe Ball, Shane Black, Danny Brown, Brian Bush, Tina Chenery, Hal and Heather Cogger, Nathan Clout, Patrick Couper, Nick Cunningham, Adam Elliott, Bailey Eipper, Cody Eipper, Ryan Francis (Ryan Francis Photography), Nick Gale, Nic Gambold, Aaron Hillier, Harry Hines, Mitchell Hodgson, Ash Horn, Hinrich Kaiser, Scott Kickham, Deb Larcombe, Brad Maryann, Michael McFadden, Danielle and Lyle McKinney-Smith, Angus McNab, Jake Meney, Brock Morris, John Mostyn, Rex Neindorf, Hamish Noller, Mark O'Shea, Dean Purcell, Wes Read, Peter Rowland, Mark Sanders (Eco-Smart Ecology), Shawn Scott, Glenn Shea, Ruchira Somawerra, Matt Summerville, Gerry Swan, Michael Swan, Steve Swanson, Janne Torkkola, Brad Traynor, Eric Vanderduys, Steve Wilson, Wolfgang Wüster, Rob Valentic and Anders Zimny. This book has been greatly improved by the constructive comments of Hal Cogger, Adam Elliott, Angus McNab, Janne Torkkola and two anonymous reviewers.

Lastly, we thank the publishers, including John Beaufoy, Rosemary Wilkinson and their staff, for the opportunity to write this work, and series editor Krystyna Mayer for ensuring readability and consistency in the text.